Science and Literature
A series edited by George Levine

Writing Biology

TEXTS IN THE SOCIAL CONSTRUCTION OF SCIENTIFIC KNOWLEDGE

Greg Myers

THE UNIVERSITY OF WISCONSIN PRESS

The University of Wisconsin Press
114 North Murray Street
Madison, Wisconsin 53715

3 Henrietta Street
London WC2E 8LU, England

5 4 3 2 1

Printed in the United States of America

Library of Congress Cataloging-in-Publication Data

Myers, Greg, 1954–
Writing biology: texts in the construction of scientific
knowledge/Greg Myers.
320 pp. cm.—(Science and literature)
Includes bibliographical references.
1. Biological literature—Social aspects. 2. Biological
literature. 3. Biology—Authorship. I. Title. II. Series.
QH303.6.M94 1990
306.4′5—dc20 89-40263
ISBN 0-299-12230-1 CIP
ISBN 0-299-12234-4 (pbk.)

for Tess and Alice

Contents

Preface

The purpose of this study is to provide some interpretations of scientific texts in their social context that will help us understand how texts produce scientific knowledge and reproduce the cultural authority of that knowledge. I shall give to drafts and published versions of biologists' writing—grant proposals, articles, and popularizations—the kind of detailed attention usually reserved for literature. But the goal of this effort is not, as it is for many literary studies, to promote an appreciation of the unique qualities of the authors or to argue for a revaluation of their writing. Instead, the close attention is meant to bring out the social aspects of scientific work, aspects that may be missed in the usual course of reading textbooks, newspaper reports, or journal articles.

For instance, among the texts I will consider are an article in *The Proceedings of the National Academy of Sciences* (*PNAS*), " 'Sexual' Behavior in Parthenogenetic Lizards (*Cnemidophorus*)," and an article in *Time* reporting the same study, "Leapin' Lizards: Lesbian Reptiles Act Like Males." One way of looking at these texts would be to evaluate them as carriers of information about a discovery: the *PNAS* article carries the information from the discoverers to the scientific community, and the *Time* article carries it from the scientific community to the larger public. But in this study I shall treat such texts, not as reporting, but as constructing scientific facts. In this view, the *PNAS* article is part of the social processes of getting other researchers to see a phenomenon of animal behavior, making a claim about this phenomenon, negotiating the place and value of this claim in the structure of scientific knowledge, and determining the place of the authors in the scientific community. The *Time* article also constructs a fact (though not, I shall argue, the same fact), places this fact in structures of knowledge ("The Sexes" section of the magazine, between "Sports" and "Cinema"), and places the zoologist in the community, as a white-coated, humorless, keen-eyed discoverer of lizard sex—an expert. It offers us the choice between accepting this new fact as "scientific," above the realm of social processes, and rejecting all such facts as trivial, divorced from common sense and common life. Either way, the *Time* article cuts us off from the social processes behind the *PNAS* article, the processes through which the fact was made.

If texts were just channels for communication of information that already existed independent of them, we could restrict ourselves to the philosophical questions about proper method and the pedagogical questions of how to teach and use the conventional forms more efficiently. But if texts are structures both for thinking and for social interaction, we can ask what they tell us about scientific knowledge. Following some sociologists of scientific knowledge, whom I introduce in the first chapter, I trace the power of this knowledge to social processes. The striking feature of scientific disciplines is the ability of their members to agree, much more than, say, literary critics, discourse analysts, or sociologists of science, on just what constitutes knowledge at any point, and what does not, and to build cumulatively on the accepted knowledge. Scientific discourse creates the consensus of the scientific community; it turns tensions, challenges, and even bitter controversies into sources of strength and continuity. Scientific texts help create the selectivity, communality, and cumulativeness that both scientists and nonscientists attribute to scientific thought.

I shall relate scientific texts to social organization and to the production of scientific knowledge by giving readings of various texts by biologists. In the first chapter I shall relate the literary approach to texts in which I was trained to some of the approaches to scientific texts I have learned from sociologists, and shall position myself in relation to some of the issues raised by these sociologists. The rest of the chapters, though they deal with several different lines of research, follow a movement through a stylized cycle of a research project, from proposals for new research, to articles reporting research to the community of specialist researchers, to controversies in the reception of these articles, to popularizations presenting specialized knowledge to a wider audience, and finally to a scientific controversy in the public forum. Each chapter shows biologists writing for a different audience: the panel of the funding committees, the editors and referees who make decisions at a journal, the members of a core set involved in a controversy, and the general public.

I start with grant proposals, the most obviously rhetorical form of scientific writing, and the genre on which all the other writing depends, because all this research depends on funding of one sort or another. The second chapter describes how two proposals are revised in the course of writing and resubmission. I shall argue that we can see in these processes how research programs that the researchers themselves believed radically challenged established ideas were incorporated into the mainstream of the discipline.

In a parallel study, the third chapter examines the refereeing and revision of two journal articles (by the same biologists whose proposals we saw in chapter 2) that had unusually bumpy rides to final journal publication. I shall argue that these processes can be seen as the negotiation of the status of the knowledge claim made in the texts.

Then, in the fourth chapter, I shall consider a controversy involving one of these researchers and other specialists in his research area, and show how the texts produced in the course of exchanges that were sometimes heated are related, as narratives that are interpreted and reinterpreted by various participants.

In the fifth chapter I consider the narratives of three popular science articles in *Scientific American* and *New Scientist*, comparing them to the authors' articles for specialized professional audiences, and looking at the sorts of changes made by the journal editors to bring the scientists' manuscripts into line with the conventions of their journal.

Finally, in the sixth chapter, I look at the construction of scientific expertise in a larger context, in the debate over the uses of scientific knowledge in society. I reconsider the debates over E. O. Wilson's book *Sociobiology* in terms of Wilson's construction of a narrative and his critics' ironic interpretation and retelling of that narrative. Where the arguments described in the fourth chapter took place at research seminars and in a few specialized journals, the sociobiology controversy takes place in the public forum, in popular magazines and newspapers, and it involves participants outside any one research group.

Like all research in the growing and rather vaguely defined academic area of science studies, this book falls between several disciplines. When I give account in the first chapter of my position between sociology and discourse analysis, I use what Nigel Gilbert and Michael Mulkay have called the "empiricist repertoire," showing how the research in scientific texts seems to lead unavoidably to just the synthesis of approaches I have presented. I could also give an account in the "contingent repertoire," of how I moved from a position as a graduate student of Victorian literature, to a job teaching scientific writing, to a job teaching literature, to a job teaching linguistics and translation, and how I moved from New York to Texas to England, and how this study developed from a simple request for abstracts to use as teaching materials for a writing course, through my own rewritten abstracts of conference papers, rejected proposals for funding, and carefully worded job application letters. All these changes in the course of writing the book leave me with the practical problem of defining (or finding) whom I'm talking to, and leave librarians with the practical problem of figuring out where this book goes in their cataloguing system.

I make an attempt to locate my approach in my subtitle, which is full of coded references to positions taken and assumptions taken for granted, aimed at potential readers in each of the disciplines to which I am appealing. *Texts* is a code word for a kind of literary study that looks beyond works of literature, for linguistic study that looks beyond the sentence, and for sociological study that sees discourse as the basis for social analysis. *Social construction* allies the book with those sociologists who see science as the product of social processes (note, for instance, Latour and Woolgar's subtitle to *Laboratory Life: The Social Construction of a Scientific Fact,* and see also their Postscript to the second edition, explaining why they changed this subtitle). Even the apparently innocent pair of words, *scientific knowledge,* signals that I will not restrict my analysis to the institutions and roles of the scientific community, as some sociologists have done, but will consider the content of science as well.

Although the jargon of my subtitle might scare some readers off, the main title is intended, not only to give some key words for the guidance of librarians, but also to suggest two broad groups of readers I hope to reach: those interested in *writing* and those interested in *biology*. I began this study when I was working as a teacher of writing and, with colleagues, was developing a writing course for natural scientists that would teach critical reading, rather than just teaching the formats and style of technical documents as given. I have tried to make it a contribution to the lively discussions about the teaching of writing, particularly of academic and professional writing, that have developed in composition programs and rhetorical studies in the United States. But I hope it retains some interest for a broader audience of those trained as scholars of literature who have applied the methods of literary criticism to nonliterary texts, and for those researchers in text linguistics, in language teaching, and in translation who have begun to look at the linguistic feature of genres.

By including the word *biology* in the title, I hope to get this book placed on the rapidly lengthening shelf of studies of science, and particularly with studies that do not necessarily take physics as the typical science. I hope it will become clear, after the first chapter, that the title is not *Writing ABOUT Biology* because that would imply that biology is there before the writing and that the writing merely dresses it up. I argue instead that writing *produces* biology. The title intentionally echoes a number of similar present-participle-plus-object constructions in titles of contemporary literary criticism and social science books (*Writing Culture, Reading Woman, Constructing Quarks*). I do not mind a cliché, and I see these titles as reflecting a focus on processes rather than on a subject.

I also hope that the title will attract biologists, though I must be cautious about what it has to offer them. It is not a how-to book, and in fact the studies in it suggest that it is very hard to give simple and unequivocal tips to those who want a book on how to write their next article. But I believe it can be useful to biologists and other scientists by pointing out which features of their texts might have rhetorical significance. As I say in the introduction, it is an important methodological question whether such a study can tell biologists anything they do not already know. I hoped at the outset that it would, that it would reveal the real science below the apparent science, like many studies of ideology in other cultural realms. But I have found that the biologists who read my chapters (all chapters were read by the writers studied and by other scientists as well) were not surprised by what I had to say, and were only surprised by the lengths to which I went to say it. Other researchers in the sociology of scientific knowledge have encountered strong resistance to their emphasis on texts from the writers they studied. But none of the scientists whose writings I examined denied that social processes were going on, and that these processes involved texts. So now I see myself, not as revealing what is hidden under biology, but as making explicit what its practitioners know, and perhaps take for granted. This is a more humble project, but one that can be useful to biologists trying to see why an article or proposal is causing them unusual difficulty, as well as to nonbiologists trying to trace the production of a scientific fact when they see only the last, public stage. That is, the project can be useful to biologists trying to get out of their assumptions, and to nonbiologists trying to get into them.

David Crews, one of the biologists whose writing I analyze, once noted, in the margin of a version of one of these chapters, "You're an ethologist." Writing researchers might take some methodological lessons from him. Like ethologists, we should not only observe and categorize the behavior of individuals, we should also consider the evolution of this behavior in its ecological context, and compare it to behavior of other species in other environments. And I agree with Dr. Crews that atypical subjects may provide the best means of reexamining received ideas about behavior. But I do not think we are ready yet for a "Natural History of Biologists' Writing." This is not just because scientists are more complicated than lizards or garter snakes, or because they won't stay in glass cages until we can perform an assay, but because our language, institutions, and authority as experts are intertwined with theirs. When Dr. Crews saw how I interpreted my data he decided I was more of a "seat of the pants ethologist." That, I think, is all one can be in the study of scientific writing today.

Acknowledgments

One of the first conventions of academic writing that I learned was that extensive acknowledgments are in poor taste. But one cannot complete an interdisciplinary project like this one without getting a great deal of intellectual and personal help. I could not have written the book without the biologists who let me read and reproduce their unpublished drafts and criticisms of their work, discussed them with me, and corrected some of my errors of interpretation. The studies that make up chapters 2 and 3 emerged from discussions with David Crews and David Bloch of the University of Texas; both were collaborators in this early stage of research as much as they were its subjects. Dr. Crews has also taken time to comment on later chapters of the book. Dr. Bloch kept me up to date with a stream of witty and thoughtful letters until his death. I miss him. Also at Texas, Lawrence Gilbert provided essential texts and commentary for chapter 5, and David Astley and Richard Friesner provided insights into biology by commenting on the writing practices of their own fields, civil engineering and physical chemistry. Since I moved to England, Geoffrey Parker and Robin Dunbar have contributed a great deal to shaping my thinking about the project, beside providing me with texts for analysis. I have also benefitted from discussions at one point or another with Eric Charnov, A. J. Cain, H. Saint-Girons, Pierre Chambon, and Stephen Green, although I have not focused on their texts in the case studies. I received written comments on my analyses of controversies from Orlando Cuellar and Charles Cole (chapter 4) and from E. O. Wilson (chapter 6). Though I am grateful for all these researchers' criticisms, and have generally incorporated them in the final text, they cannot, of course, be held responsible for what I say, for the uses to which I put their texts, or for my failures to understand biology.

I began this project as part of the rhetoric program at the University of Texas, where I benefitted a great deal from constant discussions with Lester Faigley and David Jolliffe, who were doing similar work, and from comments at one time or another by Maxine Hairston, Kristine Hansen, Susan Jarratt, and Ted Smith. I also got written coments from the farflung network of rhetoric researchers, including Ann Berthoff, Ken Bruffee, Carolyn Miller, and John Swales. My debt to Charles Bazerman's pioneering studies will be apparent in every chapter. He was the one who suggested I read some British sociologists of science; I hope he has not regretted this suggestion, since I have enthusiastically adopted their relativism. I look forward to continuing arguments at every possible occasion.

My move to England, which could have been the end of this book, instead transformed it, largely because of the welcome given by several groups of researchers. The Discourse Analysis Workshop has provided a lively and encouraging forum for these studies, even as they punctured my assumptions about texts; I am particularly grateful for criticisms from Malcolm Ashmore, Andy McKinlay, Mike Mulkay, Trevor Pinch, Jonathan Potter, Teri Walker, and Steve Woolgar. Brian Wynne, Roger Smith, and the Center for Science Studies and Science Policy helped me get in touch with the science studies group at Lancaster. Tony Dudley-Evans and his colleagues at the English Language Research Unit at Birmingham provided a linguistic forum for earlier versions of several chapters. I have benefitted from written criticisms from Martin Barker, Anne Fausto-Sterling, and Hilary Rose, who have reminded me at times of the limitations of my approach. I am also grateful to seminars in Linguistics and in the History and Philosophy of Science at Leeds, in Linguistics and in History at Lancaster, in Sociology at York, and Human Sciences at Brunel. Finally, the Modern Languages Centre of the University of Bradford enabled me to finish this book by giving me a job (with the grisly title of a New Blood Lectureship) that allowed time for research. My thanks to those who thought this area needed research, who wrote the proposal, and who pushed it through various bureaucracies.

An earlier version of chapter 2 appeared as "The Social Construction of Two Biologists' Proposals" in *Written Communication,* Vol. 2. No. 3 (July 1985), pp. 219–245, © 1985 by *Written Communication.* An earlier version of chapter 3 appeared as "Texts as Knowledge Claims: The Social Construction of Two Biologists' Articles" in *Social Studies of Science,* Vol. 15, No. 4 (November 1985), pp. 593–630. Both are reprinted by permission of Sage Publications, Inc. My thanks to the editors and referees of those journals for important suggestions. My thanks also to Barbara Hanrahan and George Levine for encouragement and responses, and to an anonymous referee for the University of Wisconsin Press for suggestions for revision. My thanks to the following for permission to reproduce copyright material: to David Crews, for permission to reproduce illustrations from an article in the

Proceedings of the National Academy of Science in appendix 3; to *Scientific American*, Inc., to the American Association for the Advancement of Science, and to Academic Press, Inc., for permission to reproduce illustrations in chapter 5; to Harvard University Press, for permission to reproduce a passage from *Sociobiology;* and to Macmillan Magazines Ltd, for permission to reproduce a table from *Nature* in the conclusion. Full copyright information is given in the illustration legends.

I owe personal debts to my parents, H. A. P. and Pat Myers, for moral support from afar, and nearer home, to Janice Mitchell, whose care for our daughter gave my wife and me time to write. My debt to Tess Cosslett goes beyond the conventional thanks to the wife, since she was the one who suggested that there was a book in this topic, and she was the one who suggested I send it to this series. She has also endured many hours of dinner table conversation on such topics as the Strong Programme, evolutionarily stable strategies, and garter snakes. Alice's contribution was less direct, but this book would never have been finished if she did not now and then, at last, fall asleep.

Writing Biology

Chapter One
Controversies about Scientific Texts

Researchers in philosophy, sociology, social psychology and anthropology have recently become interested in the discourse of their own and other disciplines.[1] Words such as *text, discourse, narrative,* and *construction* have become fashionable, and have taken on a number of different, and perhaps inconsistent, meanings. I shall try to place myself in this field of research, with its rapidly shifting disciplinary boundaries, by addressing at the outset two key questions: Why study *scientific* texts? Andy why study scientific *texts*? In each answer I draw some flexible, perhaps paradoxical, demarcations between the scientific and nonscientific, between text and praxis; other researchers, as I shall show, draw the lines in different places. I will not try to insist here that my approach is correct or even consistent, only that it is methodologically practical.

The questions about the value of studying a *scientific* texts could come from two different groups of people: (1) those who believe scientific knowledge must have a special status, so that scientific texts are, at least in their ideal form, exempt from rhetorical or literary analysis, and (2) those who see scientific knowledge as having no special status, so that the only goal of a study like this one can be to

1. See, for instance, on philosophy, Jonathan Ree, *Philosophical Tales* (London: Methuen, 1987); on biochemistry, Nigel Gilbert and Michael Mulkay, *Opening Pandora's Box* (Cambridge: Cambridge University Press, 1984); on history, Hayden White, *Tropics of Discourse* (Baltimore: Johns Hopkins University Press, 1978), and *Metahistory* (Baltimore: Johns Hopkins University Press, 1973); on anthropology, George Marcus and Michael M. J. Fischer, *Anthropology as Cultural Critique: An Experimental Moment in the Human Sciences* (Chicago: University of Chicago Press, 1986); on psychology, Michael Billig, *Arguing and Thinking: A Rhetorical Approach to Social Psychology* (Cambridge: Cambridge University Press, 1987), and Jonathan Potter and Margaret Wetherell, *Discourse and Social Psychology* (Beverly Hills and London: Sage, 1987); John S. Nelson, Allan Megill, and Donald McCloskey, eds., *The Rhetoric of the Human Sciences: Language and Argument in Scholarship and Public Affairs* (Madison: University of Wisconsin Press, 1987); James Clifford and George Marcus, eds., *Writing Culture* (Berkeley: University of California Press, 1986); Michael Lynch and Steve Woolgar, eds., *Human Studies: Special Issue on Representation in Science* 11 (July 1988).

show that scientific texts can be treated in the same way as literature or political oratory. Those who believe that scientific discourse is essentially different from other discourse—including some realist philosophers of science, some Marxists, and some practising scientists, at least in their polemical statements—point to a distinctive scientific method involving falsification or replicability, to institutions such as peer evaluation and publication, to the position of the scientist in historical processes, or to some quality of the subject matter studied. These distinctive characteristics of science are taken to separate science from the realm of rhetoric and of social processes, so that, however much social factors may enter in any particular case, *real* science always continues, or works best, apart from those factors. But as I shall show, the application of this demarcation is itself a question of rhetoric and social processes; such characteristics as replicability are invoked in order to persuade the audience that some fact or field lies beyond matters of persuasion. Science is like other discourses in relying on rhetoric; it just uses a different kind of rhetoric. Traditional literary critics draw the same sort of demarcation between what can and cannot be studied as discourse, but they draw it from the other side, resisting the application of literary criticism to anything but literary texts.

If science relies on rhetoric, it might seem that I could subsume this study under general studies of discourse formations. It would be convenient for those of us trained in textual analysis if all discourses could be reduced to one discipline, preferably our discipline, so that I would have completed my task if I could find in these texts the major tropes of Northrop Frye's *Anatomy of Criticism* or the categories of Aristotle's *Rhetoric*. But such a project, even if it were successful, would not help to explain why, in our culture, scientific knowledge has a huge authority, and literary criticism, for instance, does not. I shall argue (following some sociologists) that an understanding of the discourse of any discipline depends on a detailed knowledge of that discipline—not just a knowledge of its content, since the construction of that content is what is at issue, but a knowledge of its everyday practices. For my purposes, the crucial difference between articles by a psychobiologist and a literary critic lies not in some quality of the subject matter, not in the fact that one is writing about garter snake hormones and the other is writing about John Ruskin's symbols, but in the form of the article and the kind of rhetoric it allows. The psychobiologist, for instance, can make use of the work of ethologists, ecologists, and chemical assay designers to support his or her claim, whereas the literary critic relies on his or her own authority, and is likely to invoke other critics mainly to challenge them. Bruno Latour makes this point.

> Rhetoric used to be despised because it mobilised *external allies* in favour of an argument, such as passion, style, emotions, interest, lawyers' tricks, and so on. . . . The difference between the old rhetoric and the new is not that the first makes use of external allies which the second refrains from using; the difference is that the first uses only *a few* of them and the second *very many*.[2]

This view of scientific rhetoric is uncomfortable for the nonscientist studying scientific texts; to follow the scientist, as Latour suggests, one has to know about all these possible allies, and about the ways they can be invoked. I cannot become a biologist, but I do focus on just a few areas of research, so that I can deal in some detail with the practices of those specialties.

The question of why one should study *written* texts is raised by science studies researchers who believe that the reliance of historians of science on the published literature has led them away from the actual practices of science. The limitations of what Michael Lynch calls "literary" analysis have been pointed out, for instance, by Peter Medawar, Harry Collins, and Bruno Latour, who says,

> No matter how interesting and necessary these studies are, they are not sufficient if we want to follow scientists and engineers at work; after all, they do not draft, read and write papers twenty-four hours a day. Scientists and engineers invariably argue that there is something behind the technical texts which is much more important than anything they write.[3]

These researchers advocate such techniques as ethnography, participant observation, and conversation analysis to get behind the written texts. Certainly studies of the talk and actions of scientists by these and other researchers are crucial for an understanding of science and of scientific writing. But written texts have great advantages as re-

2. Bruno Latour, *Science in Action: How to Follow Scientists and Engineers Through Society* (Milton Keynes: Open University Press, 1987), p. 61.

3. Latour, *Science in Action*, p. 63; Michael Lynch, *Art and Artifact in Laboratory Science: A Study in Shop Work and Shop Talk in a Research Laboratory* (London: Routledge and Kegan Paul, 1985), pp. 143–54; Peter Medawar, "Is the Scientific Paper Fraudulent?" *Saturday Review* 49 (1 August 1964): 42–43; Harry Collins, *Changing Order: Replication and Induction in Scientific Practice* (Beverly Hills and London: Sage, 1985), p. 73.

search material, advantages that have long been taken for granted by literary critics, but are perhaps not sufficiently appreciated either by them or by social scientists:

1. Texts hold still.
2. Texts are portable.

In this study, I want to make a virtue of the necessity of converting material into a written form. Because all the materials I use are written, even the informal comments in the margins of drafts, I can reread them slowly, again and again, display fragments, and read back through them to find other instances of a feature I've just noticed. Because I can use anything written, and need not choose and transcribe special moments, I have an endless stream of material in the overflowing filing cabinets of the scientists I study. This availability for close reading, and this wealth of material, means my reading of scientific texts is not like the normal reading processes of the scientific writers, referees, or readers themselves. But the strangeness of my position, as a literary reader of scientific texts, allows me to bring out features that otherwise pass unnoticed. Only written texts allow for such close reading, because they hold still while one goes over them, and hold still until one can come back to them.

The other advantage of written texts as research material is related to this: texts, unlike conversations or experience, are portable. When Michael Lynch studies a conversation about an electron micrograph between a lab director and a postdoctoral researcher, he is dealing with something local. He can make it available for discussion only by taping it and transcribing it according to conversation analysis conventions, transforming it into a written text that is an idealization of the fleeting moment and complex interaction that Lynch wants to discuss. When I quote a sentence as an example, it can pass from the author's word processor, though a photocopy machine, into my word processor, and into the typeface of this book, all, for my purposes, unchanged. When I quote a published text, anyone can go to a library and look up the same article in another copy of *Nature* or *Sociobiology*. A reader and I can argue about the same thing; the reader and author of an ethnography do not both have access to the same experience. There is certainly a rhetorical advantage is being able to point to an example and say, "There it is in black and white."

It may seem paradoxical that I would defend my use of written texts on the grounds that texts hold still and texts are portable, when in every chapter of this book I shall be arguing that texts must be read as processes, not objects, and that texts change meaning whenever they change context. Surely, then, the sentence I point to as evidence does not hold still, and is not portable. In literary scholarship, bibliographical scholars could point out that the texts of canonical works hardly hold still, but change with each generation, and reader-response critics could point out that the meanings of works are not really portable between different contexts. These questions are all ways of calling attention to the processes of production and interpretation; I shall be stressing those processes but shall still draw on the practical advantages of words on the page. It remains true that the way one makes an argument in literary criticism, whatever one's approach, is to quote a fragment of a larger text, trusting that the properly guided reader will have the response the critic needs, and that this response will stand in for all the processes the critic is trying to bring out. I can make my sociological argument convincingly, opening up the processes of texts and showing the diversity of interpretations, only because written texts can function as evidence on this basic level. I use written texts, not because I hold them to be in any way primary, but for the practical reason that I can do things with them that I cannot do with other data.[4]

My answers to both these questions—why study *scientific* texts and why study scientific *texts*—are made in relation to the assumptions of someone trained in literary analysis. To approach the first question, I have to reject the assumption that literary and nonliterary texts are essentially different, while recognizing that the practices of science could be different from the practices of other discourses. To answer the second question, I have to make explicit the practical advantages of written texts in arguing for a view of science. Most of my references in this book will be to sociologists of science, but I find, rather to my surprise, that this book is still a literary and rhetorical study. It is literary not because it responds to the latest approaches in literary theory (it does not), but because it uses the skills and draws on the assumptions of someone trained in literature, rather than in the sociology of science or evolutionary biology. How, then, does my literary approach relate to those of researchers drawing on other disciplinary assumptions?

In the following sections, I examine examples, first, of literary and

4. My argument for the methodological advantages of written texts as sources forthe sociologist of science parallels the argument for their importance in the history of science. The classic presentation of this relation is by Elizabeth K. Eisenstein, *The Printing Press as Agent of Change* (Cambridge: Cambridge University Press, 1979). Bruno Latour has developed it into the concept of "immutable mobiles"; see *Science in Action*, p. 223.

then of sociological approaches to science. If one considered only the research done before the 1970s, literary critics and sociologists would not seem to have much in common in their approaches to scientific texts. Literary criticism, though it allowed the study of science in relation to literary texts, as part of the cultural background, seemed to exclude any consideration of scientific texts in relation to science. And sociologists of science, though they were very interested in scientific communication and the institutions around publication, didn't seem to be interested in reading individual texts. In both cases the lack of interest in scientific discourse, the exclusion of the nonliterary from literature and of the noninstitutional from sociology, were not incidental oversights, but helped to constitute both disciplines, allowing them to ask their characteristic questions and evaluate the answers.

In examining the sociology of scientific knowledge, I focus on some persistent controversies that deal with the two questions I have already raised, about the demarcation between science and nonscience, and between text and practice. This review does not provide a broad and balanced introduction to the sociology of scientific knowledge,[5] but points out some of the tensions that will surface again and again through this book. I see each of the traditional approaches to scientific texts, through literature, history, or sociology, as avoiding questions of the relation between knowledge and its textual representation. As discourse analysts in various disciplines have shown, to challenge such exclusions is not to expand the methods of literature or history or sociology into some new material, but to transform the discipline.

Traditional Literary Approaches to Scientific Texts

I can illustrate the usefulness and the limitations of traditional literary approaches to scientific texts by considering a 1968 essay by Dwight Culler, "The Darwinian Revolution and Literary Form."[6] Culler's

5. For introductions to the Strong Programme, see Michael Mulkay, *Science and the Sociology of Knowledge* (London: George Allen and Unwin, 1979), a readable argument for the general reader; Barry Barnes and David Edge's anthology, *Science in Context: Readings in the Sociology of Science* (Milton Keynes: Open University Press, 1982), a selection of important studies; Mary Hesse, *Revolutions and Reconstructions in the Philosophy of Science* (Brighton: Harvester, 1980), a philosophical defense; and Steven Shapin's massive review article, "History of Science and Its Sociological Reconstructions," *History of Science*, 20 (1982): 157–211.

6. Dwight Culler, "The Darwinian Revolution and Literary Form," in *The Art of Victorian Prose*, ed. George Levine and William Madden (New York: Oxford University Press, 1968).

work is appropriate for showing some of the assumptions in the field and the craft skills that are admired in its practitioners. One might think, from the ferocious battles in *Critical Inquiry* or other journals, that literary criticism was an especially heterogeneous and contentious discipline. But while the theory of literary criticism is fragmented into competing schools, there is a powerful consensus in the practice of literary criticism, especially in the daily work of teaching and evaluation. Lacanian and Leavisite, Derridian and Alhusserian show remarkable agreement in distinguishing an upper second class and lower second class examination paper (or, in the United States, an A− from a B+ student). Literature has a notion of craft skills more narrowly defined but less explicitly articulated than that of, say, linguistics or sociology. Culler offers a fine example of those craft skills. And I too employ those skills, though I do not employ them so well.

I have chosen this relatively dated essay from the huge bibliography on literature and science, not only because it is a fine essay, perceptive, witty, and surprising in the connections it draws, but also because it can be seen as the forerunner of later studies that relate science to literary form.[7] Culler notes that most earlier studies of the influence of Darwin on literature show how some ideas related to his evolutionary theory are treated in works of literature. In the kind of turn from the thematic to formal analysis characteristic of criticism of the 1950s and 1960s, he sets out instead "to inquire how the form of Darwinian explanation has influenced, or is analogous to, forms of literary expression in the post-Darwinian world" (p. 225).

In Culler's argument, "the form of Darwinian explanation" is the reversal of Paley's argument from the evidence of design in nature to the existence of God the designer. Darwin "has abandoned the teleological explanation, which looks to the future, for a genetic explanation, which looks to the past. . . . Where Paley has taken intelligence to be the cause and adaptation to be the result, Darwin has shown

7. Influential studies of science and literature are Gillian Beer, *Darwin's Plots: Evolutionary Narrative in Darwin, George Eliot, and Nineteenth-Century Fiction* (London: Routledge and Kegan Paul, 1983) and Sally Shuttleworth, *George Eliot and Nineteenth-Century Science: The Make-Believe of a Beginning* (Cambridge: Cambridge University Press, 1984). My own articles, "Nineteenth-Century Popularizations of Thermodynamics and the Rhetoric of Social Prophecy," *Victorian Studies* 29 (1985): 35–66, and "Science for Women and Children: The Dialogue of Popular Science in the Nineteenth Century," in *Literature and Science 1700–1900*, ed. Sally Shuttleworth and John Christie (Manchester: Manchester University Press, forthcoming), are typical examples of work in this area. The collection edited by Shuttleworth and Christie contains a number of historical studies.

that adaptation was the cause and survival the result—survival of those fittest to survive" (pp. 227–28). Culler compares this reversal to those performed by Malthus, Bentham, and Hume in their arguments with earlier thinkers. He finds, because of this juxtaposition of the pompous figure and the ironic questioner, "a fundamental analogy between the Darwinian explanation and the whole comic, satiric tradition" (p. 237). The entertaining turn to Culler's essay is the application of this formula to a whole range of literary works, from those that might be expected in an essay on Darwin's influence—*Erewhon* and *Man and Superman*—to those that are a surprise in this context—*Alice in Wonderland,* Pater's *Renaissance,* and *The Picture of Dorian Gray*. All, Culler argues, share this pattern of reversal, this characterization of fixed old ideas and disruptive new ideas, this confrontation of explanation in terms of design with explanation in terms of chance.

Culler's performance is a good example of the procedures literary critics take for granted. He focuses on matters of form, selecting a few telling features, organizing them into a pattern, and taking them to define whole texts. He uses comparison and juxtaposition to make these formal features stand out. He is interested in the use one text makes of another text, and in the possibilities for reinterpretation in these juxtapositions. He identifies texts with the authors as represented in the text, and imputes to these authorial personae various intentions and interests. In all this, Culler's article exemplifies the procedures I will be following in this book.

Culler also exemplifies some habits from academic literary study that I am trying to break. He completely ignores Darwin's text and its context in scientific discourse. His article is typical of literary studies in tracing influence only in one direction—from science to literary texts. For all its broad range of literary erudition, it refers to no other scientists beside Darwin. The questions of what conventions Darwin was following, what influences he might have felt, what rhetorical purposes he might have had, are not raised, the way they would be raised with any literary author. The scientific works Culler does refer to are all part of a literary canon; his selections are interesting, but all of them could be in a literature course on the Victorian or modern period. This kind of relevance seems to be necessary to justify excursions into the scientific literature. More recent studies of science and literature give greater attention to scientific discourse, but even in the excellent work of, say, Gillian Beer, the ultimate goal is to enrich our understanding of works in the literary canon. Very rarely do literary critics use their skills to help us understand science.

Like Culler and most literary critics, I reject the assumption of some philosophers and historians that the analyst can abstract meanings from texts, and take these disembodied "themes" as the objects of discussion. But Culler has so little interest in the scientific literature, as opposed to canonical literature, that in an essay on form he does not need to refer to Darwin's text at all. "The form of Darwinian explanation," as it turns out, is not the narrative structure of *The Origin of Species;* it is the relation that Darwin as a figure has to Paley as a figure. Culler gracefully notes in passing that the text does not, in fact, illustrate the form he attributes to Darwin.

> As a scientist he was the author of the greatest repartee in nature, but as a man he says he was without wit and that he had a fatality for putting his statements initially in a wrong or awkward form. Surely this is borne out by the form of the *Origin of Species.* Who of us, if we had the opportunity to write such a book, would not begin, as in a drama, by building up Paley and his argument by design with the whole range of existing plausible fact and then, by a quick reversal, bringing it all tumbling down with an explanation so simple and obvious that Huxley would slap his knee and say, "Why didn't I think of that?" and others would wonder and find the new view as satisfying as it was surprising? I do not say that this is the way to get the theory accepted, but simply in order to present it, as a brilliant and paradoxical theory, this is the way. (P. 232)

Culler also gives Darwin an epigram that he "might have said," and comments in a note that "I am not referring to the historical Darwin who in successive editions adopted a compromise position, but to an abstract 'Darwinism' which says sharply what he ought to have said" (p. 246n).

I enjoy the high-handedness with which Culler brushes aside the text that refuses to fit, but I also see a missed opportunity. It is not that this critic, one of the best analysts of Victorian prose, cannot analyze Darwin, but that somehow Darwin, a scientist, falls outside of his area of interest, where John Henry Newman, for instance, does not. I would argue that he cannot include Darwin without undermining the goals of traditional literary analysis, that is, appreciation and evaluation. The terms with which he dismisses Darwin are significant. In his view, Darwin is not a particularly good writer. Darwin should have found a better form to fit the aesthetic appeal of the theory, even if this form would not have fit his rhetorical purpose. Culler denies any interest in what form might have persuaded someone in that particular context to accept the idea; he instead imagines a context-free presentation that would bring out the qualities admired by literary critics. Clearly it is necessary for such criticism to maintain an independent aesthetic realm, in which works from different contexts can all be evaluated, as for instance when Culler compares *Erewhon* to *Gulliver's Travels* at the end of his essay. Explicitly rhetorical texts suggest another standard of evaluation—the *Origin of Species* is a successful book because it worked, within the discourse of mid-Victorian biology. And any study of the discourse of which it was a part suggests a whole set of interrelations among texts that are not like those of a literary tradition, interrelations that lead us to the immediate context rather than to the timeless.[8] So this exclusion is not an oddity of Culler's, but is part of the structure of the field. If one breaks down this barrier, one cannot do the same kinds of evaluation. One can make subtle, close readings of scientific texts, but they always have to lead to literary texts and literary questions.

If issues in the analysis of scientific texts emerge most clearly in controversy, the issues involved in an analysis like Culler's can be seen most clearly by comparing it to a historical study in the same collection, Walter Cannon's brilliant, perverse essay on "Darwin's Vision in *On the Origin of Species.*" It is brilliant in the way in which the historian Cannon performs the literary analysis that the literary critic Culler did not: analyzing the language and structure of the text in the content of the genre as defined in its period. Cannon comments on the form of treatises in the nineteenth century, compiles and compares words Darwin used in his descriptions of several other scientists, relates the rhetoric of his paragraphs to an unconventional notion of scientific logic, and explains the function of each chapter in the development of the book as a whole. Cannon is not, in fact, responding to Culler; he and Culler are both responding to what they see as the limitations of a famous and rather crude essay by Stanley Edgar Hyman. But Cannon's sarcastic swipe at Hyman could apply equally well to any literary critic who writes about scientific texts without considering their scientific and social context: "Perhaps my essay should be read as a test case as to whether a close reading of the given text, and a historical knowledge of the period in question, are useful tools in literary analysis" (p. 154).

Cannon's criticisms of Hyman and of literary criticism in general hit home, but the essay is perverse because the careful analysis of scientific prose leads only to the conclusion that the analysis of scien-

8. Robert Young shows this in his essays in *Darwin's Metaphor* (Cambridge: Cambridge University Press, 1986).

tific prose is not worth doing. For Cannon, as for many historians, the ideas of scientists can be considered apart from any textual representation, so specialists in textual analysis are left with nothing to do with science.[9] Here, ironically, he comes close to agreeing with Culler.

> Scientifically, the *Origin* is a classic; biologists have been scrambling for a hundred years to catch up with Darwin's ideas. But verbally it is a rag-and-bone shop. Science took wings in the middle of the nineteenth century, imaginative wings that no other discipline can match even now. Words, logic, evidence, and mathematical consistency tend to strangle a scientist's idea before it can ever be born alive and gasping. . . . In the *Origin* Darwin sees that it is the *habit of looking at things in a given way* which a master scientist transmits to his disciples. How he does this, rhetorically, is of little importance. If this means that literary criticism as practised by professors of English literature at liberal arts colleges is not of much importance in understanding the important books of the world since 1859, I am sorry. (Pp. 172–73)

Cannon and Culler agree that Darwin is not a good writer, and that his ideas can be considered apart from their textual form. In part they are both reacting to an analysis like Hyman's that would focus on a few elaborate and atypical paragraphs as the rhetoric of the work. But they are also defending the cores of their respective disciplinary approaches. In praising Darwin the scientist and dismissing the writer, Culler preserves a realm of the aesthetic. In the same way, Cannon preserves a narrative of the progress of science in a realm of ideas apart from the social and the textual. To grant that these ideas exist only in their textual form would be to tie science to the limited culture of a particular period. Just as aesthetic criteria are the basis of Culler's conclusion, the narrative of progress is the basis of Cannon's, so the

9. For examples of historical work that does focus on textual issues, see Frederick Holmes, "Lavoisier and Krebs: The Individual Scientist in the Near and Distant Past," *Isis* 75 (1984): 131–42, and "Writing and Discovery," *Isis* 78 (1987): 220–35; Jan Golinski, "Robert Boyle: Skepticism and Authority in Seventeenth-Century Chemical Discourse," in *The Figural and the Literal*, ed. Andrew Benjamin, Geoffrey Cantor, and J. R. R. Christie (Manchester: Manchester University Press, 1987), pp. 58–82; and Martin Rudwick, *The Great Devonian Controversy: The Shaping of Scientific Knowledge among Gentlemanly Specialists* (Chicago: University of Chicago Press, 1985). Steven Shapin argues that historians of science have never neglected the issues raised by discourse analysts in "Talking History: Reflection on Discourse Analysis," *Isis* 75 (1984): 125–28, his response to Nigel Gilbert and Michael Mulkay's "Experiments Are the Key: Participants' Histories and Historians' Histories of Science," *Isis* 75 (1984): 105–25.

Origin is good because "biologists have been scrambling for a hundred years to catch up with Darwin's ideas."

I use the methods of both the literary critic and the historian in this book, but I am trying to answer a question that is not addressed by either of their approaches—in what way do texts contribute to the social authority of science? What I need are not different methods, but a different disciplinary orientation—I need to look at studies that can lead from texts to the social structures of science, not always from scientific ideas to literature, or from scientific texts to disembodied scientific ideas.

Sociological Approaches to Science: The Mertonian Paradigm vs. the Strong Programme

One reason the sociology of science seems to have little to do with literary analysis is that it was long dominated by a paradigm that leaves the analyst of texts with little to do. And yet the creator of this paradigm, Robert Merton, raised many of the questions about the cultural authority of science that I am trying to raise here. He challenged the popular sense that science involves the confrontation of the individual scientist with nature and so prepared the ground for work like mine. A brief summary of the form in which he put his questions may help show the difference between the approach I am following (once largely British) and that of much of the sociology of science done in the United States.

Merton and his students developed, starting in the 1940s, a framework for analyzing the function of scientific institutions as the harnessing of the motives of individuals to serve the ongoing interest of the scientific community as a whole. The overriding interest of this community was in extending validated knowledge. Within Merton's framework there has been a vast literature on how institutional factors such as the reward system or hierarchies of status serve the progress of science. The Mertonian paradigm seems to allow sociology of science to become like a science, and both Mertonians like Norman Storer and critics like Barry Barnes agree on its influence in shaping the development of the discipline.[10]

10. For a review citing major articles in the British vs. American debate of the early 1970s on the Mertonian paradigm, see Charles Bazerman, "Scientific Writing as a Social Act: A Review of the Literature of the Sociology of Science," in *New Essays in Technical and Scientific Communication: Research, Theory, and Practice,* eds. Paul V. Anderson, R. John Brockmann, and Carolyn R. Miller (Farmingdale, N.Y.: Baywood, 1982).

One Mertonian study that deals with texts is an article by Harriet Zuckerman and Merton, "Patterns of Evaluation in Science: Institutionalization, Structure, and Functions of the Referee System" (collected in Merton's *The Sociology of Science*). For Zuckerman and Merton, texts are an important part of the institutions of science and can be studied statistically in the form of ciatation data and bibliographical records; nonetheless, but they are treated as if they are just vehicles for the communication and validation of technical knowledge, not in themselves shapes of that knowledge. Zuckerman and Merton separate one part of science that is social, and subject to analysis in terms of institutionalization, structure, and function, from another part of science that is technical and that relates to the natural world. The interpreter approaches the texts objectively, as data to be tabulated, rather than as part of discourse to be interpreted. I shall try to show that this framework limits the analyst of texts who wants to ask about the cultural authority of science. It seems to me that the end result of Merton's approach in this article is a defense of science and the uncritical adoption of its methods for sociology.

The first part of Zuckerman and Merton's article is a fascinating brief history of the earliest years of the refereeing system, at the founding of the *Philosophical Transactions* of the Royal Society. Merton, who did pioneering historical work on the social background of seventeenth-century science, is able to give a historical perspective which sociology sometimes lacks. The assumption that science starts in the seventeenth century is important to Zuckerman and Merton's approach. They can trace refereeing back to its origins because the institutions of science are part of a structure that carries through many periods and cultures, changing but retaining its identifying characteristics. This structure comes into being when it becomes institutionalized. For instance, they argue that "these institutions provided the structure of authority which transformed the mere *printing* of scientific work into its *publication*" (p. 402). These institutions are analyzed in terms of their systematic functions. "As with the analysis of any case of institutionalization, we must consider how arrangements for achieving the prime goals—the improvement and diffusion of scientific knowledge—operated to induce or to reinforce motivations for contributing to the goals and to enlist those motivations for the performance of newly developing social roles" (p. 464). For instance, the desires of individual researchers for recognition and for protection of their intellectual property rights, and the desires of individual readers for the sharing and prior assessment of knowledge, were both served by an evolving system of refereeing, and this system served the accumulation of certified knowledge that is Merton's science.

After this historical vignette, Zuckerman and Merton go on to a series of studies more typical of other Mertonian articles—highly refined statistical analyses that relate patterns in some body of data to the functioning of some institution of science. So, for instance, they compare rejection rates of journals in various academic fields, find a pattern that the rejection rates are, in general, far higher in the humanities and social sciences than in the natural sciences, and relate this pattern to the degree of institutionalization of these fields. "This suggests that these fields of learning [in which many manuscripts fail to meet minimum standards] are not greatly institutionalized in the reasonably precise sense that editors and referees on the one side and would-be contributors on the other almost always share norms of what constitutes adequate scholarship" (p. 472). So this statistical pattern can lead to a demarcation of science from nonscience. The implication, as in the historical study, is that the institution is functional; the shared norms keep scientists from wasting their time on studies that will not be rewarded and will not further the accumulation of knowledge.

The rest of Zuckerman and Merton's article focuses on data from the *Physical Review* concerning the acceptance or rejection of articles, in relation to the status of the authors and of the reviewers, with this status assigned in terms of a three-tier hierarchy based on professional awards. So, again, the problem is to relate observable data to the functioning of a crucial scientific institution. For instance, they show that there was no tendency to give the manuscripts of high-ranking physicist authors to high-ranking physicist reviewers, nor were there other statistical patterns of bias in assignment. From this they suggest, albeit tentatively, "that expertise and competence were the principal criteria adopted in matching papers and referees" (p. 485). Similarly, they show that high-ranking reviewers did not reject low-ranking authors or favor high-ranking authors more than other reviewers did, nor were there any other patterns indicating bias in evaluation. "This suggests that referees were applying much the same standards to papers, whatever their source" (p. 491). The article concludes with a section discussing how the evaluation procedure allows science to progress by assuring that "much of the time scientists can build upon the work of others with a degree of warranted confidence. It is in this sense that the structure of authority in science, in which the referee system occupies a central place, provides an institutional basis for the comparative reliability and cumulation of knowledge" (p. 495).

Since I offer this article as an example of what I shall *not* be doing with texts, I should in fairness acknowledge its relative strengths. In its careful relation of statistics to institutional structures, and institutional structures to functions, Zuckerman's and Merton's article responds to sociological criteria for explanation, and it clearly produces information, and certainty about that information, that my own study of the referee system in chapter 3, with just two cases, could not. The complexity and subtlety of the Mertonian system is apparent in the description of a tension between the hierarchy that defines the structure of the scientific community and the norms that define its ethos. The norms of science, the most controversial part of Merton's model, require that scientists apply universal standards, share their knowledge, remain disinterested, and approach claims with organized skepticism. There is a tension between these norms and the hierarchy in which some scientists are regarded as better, as more worthy of being listened to and believed, than others. The review procedure, as Zuckerman and Merton describe it, offers an institutionalized means of reconciling this tension, so that the existence of a hierarchy does not distort the normative behavior on which progress depends. As Zuckerman and Merton point out in a note, their view is rather more subtle than either the view that the scientists make judgments based solely on scientific criteria, or the view that scientists make judgments based on their status in the system.

But for all this subtlety and informative power, the approach of this article is remarkably unhelpful for the question that I have set out to answer: how do texts construct scientific authority? First, Zuckerman and Merton do not look at any texts. This is partly because they can produce more convincing claims with a large-scale study, covering hundreds of articles, that precludes analysis of individual items. But I would argue that their approach would not, in any case, lead one to texts. When they suggest an experimental design that might control for "papers of the same scientific quality" (p. 486), they seem to have in mind texts as simple entities that can be assigned a value. When they sum up a broad corpus of articles on all areas of physics, articles written in a variety of contexts, they seem to assume that the content of the articles does not matter. When they categorize responses of referees simply in terms of rejections, acceptances, and the category of "problematic manuscripts" for those that may be accepted if revised, they assume that what the referees' comments say and how they say it, apart from their decisions, do not matter. It is not that I would require Zuckerman and Merton to do the study I have done—I'm glad they did not—but that I don't see their approach as following from the same assumptions as mine, even when we are looking at the same sort of materials. Actual texts and specific disciplinary knowledge are, in their approach, the noise that needs to be filtered out for the signal to come through clearly.

Also, Zuckerman and Merton make it very difficult to pose questions about scientific authority. Science is to be explained, not by analysis of the work of scientists, but by showing how scientific institutions produce science. The assumption is that there is such a thing as objective scientific knowledge, the accumulation of which may be furthered or hindered by various societies—furthered by seventeenth-century England, or hindered by Nazi Germany—but the propositional content of which is independent of any society. In the study of article refereeing, if they can show there is no statistical evidence of bias in decisions, then those decisions must be based on purely scientific "expertise and competence" or "standards." But the dichotomy of bias and objectivity makes it impossible to look at a social process in science that is neither an objective encounter with natural fact nor a dishonest departure from the fact. As the other sociologists I shall discuss show, the social *construction* of a scientific fact doesn't fit either of these categories. Remarkably often, other Mertonian studies, like Jonathan Cole and Stephen Cole's massive study of the grant review procedure, conclude that scientific institutions are most likely to produce scientific knowledge. But what if scientific knowledge is merely that which is produced by scientific institutions? This circular framework of argument grants the legitimacy and independence of scientific authority at the outset, when it is just the production of that authority that I want to investigate.

The Zuckerman and Merton or Cole and Cole studies constitute a polemic in favor of the unhindered functioning of scientific institutions. My disagreement with this polemic is political as well as methodological. In Merton's work in the early 1940s the need for the unhindered progress of science was one argument for democracy.[11] But it can be argued that today the unquestioned authority of science can itself be a danger to democracy. In the terms which Merton uses—bias or objectivity—science always remains unquestioned because it corrects its own errors. I would argue that closer study, although it will not show bias in the referee system, shows that the sense of the

11. See his influential 1942 essay, "Science and Technology in a Democratic Social Order," collected in *The Sociology of Science.*

objectivity of science is part of what science produces in scientific texts. So the objectivity of science is not an argument for giving scientific experts special authority in the political process.

Some of these limitations of the Mertonian paradigm are overcome in the recent work of Charles Bazerman, who takes the paradigm as the basis for his studies of scientific texts. In his 1984 article "Modern Evolution of the Experimental Report in Physics" (reprinted in *Shaping Written Knowledge*), Bazerman, like Merton, provides a broad-based study of changing scientific institutions. He selects representative articles from the *Physical Review,* now the most important journal in physics, focusing on those in the field of spectroscopy. But Bazerman approaches these articles in two ways: "using [stylistic] statistics to indicate gross patterns or trends but using close analytical reading to explore the finer texture, the meaning and the implications of those trends. The statistics indicate that something is happening, and the close readings are to find out what that something is" (p. 167). In the statistical portion of the paper, Bazerman relates the increasing length and the increasing number of references to increasing knowledge and the tighter links between work done in the field. He asserts that changes in clause structure and word choice reflect changes in argument. The changes in graphics and structure of articles also reflect theoretical integration. Bazerman discusses a chronological sequence of selected articles "to suggest a rhetorical history of the field" (p. 184). These close readings show in detail the growing awareness of the theoretical framework behind any experimental report.

I would argue that Bazerman's article goes far beyond the approach of Zuckerman and Merton. First, he is concerned with texts, which, as I have noted, are noticeably missing from Zuckerman and Merton. His analysis of citation data benefits from the increasingly close analysis of context and content of citations by Daryl Chubin and others.[12] Also, he limits his corpus to one field, so that he can make comments about the content of articles. For instance, he says of one early article. "The data presented are not selective concerning an issue at hand, but rather seem presented for their own sake" (p. 185), something he would not see if he had no awareness of the context of argument in the field. One may criticize details of Bazerman's article; for instance, the grammatical categories he uses are rather crude for the highly specific interpretations he wants to make of the results. But he tries to persuade textual scholars to look at social systems, and he tries to

12. See, for example, John Swales, "Citation Analysis and Discourse Analysis," *Applied Linguistics* 7 (1986): 39–56.

persuade sociologists to look at texts.[13] "As much as any of the other institutional arrangements of science, writing conventions are significant social facts for the communal operation of science" (p. 191). Bazerman concludes this study by observing that "The large-scale trends revealed here are consistent with the traditional view that science is a rational, cumulative, corporate enterprise, but point out that this enterprise is realized only through linguistic, rhetorical, and social means and choices, all with epistemological consequences" (p. 191). I do not think his work—or mine—need necessarily support this traditional view of science.

For my questions about the social authority of science, I have found the most useful guides in some of the studies that have led away from the Mertonian approach toward a model based on indifference to claims of truth, on the importance of the socially contingent, and on conflict. Barry Barnes named this approach the "Strong Programme" in the sociology of knowledge because it would find a social basis for all knowledge, not just for certain irrational beliefs. The key issue on which researchers in the Strong Programme have broken with the Mertonian tradition is the assumption that the content of natural knowledge can be separated from the social processes that produce it. Merton shows how institutions function to further the accumulation of knowledge about the natural world. The researchers in the Strong Programme would argue that the knowledge itself can be seen as social; they distinguish their field from that of the Mertonians by calling it the sociology of scientific *knowledge*. One way these sociologists bring science into the realm of sociology of knowledge is by arguing that explanations of beliefs should be symmetrical; that is, one should not distinguish between "correct" and "incorrect" beliefs in making explanations. One should use the same modes of explanation for belief in witchcraft or phrenology as for belief in electromagnetic waves or neuroendocrinology. The particular explanations behind these beliefs may, of course, be different, but one can't say, in this approach, that the nineteenth-century public believed in phrenology for cultural rea-

13. Bazerman's studies have been useful to such applied linguists as John Swales and Tony Dudley-Evans, who are trying to put linguistic research on scientific texts, and the teaching of English for Special Purposes, back in a social context. (For Swales. see n.12; Dudley-Evans has edited "Genre Analysis and ESP," *ELR Journal* 1 [1987], available from the English Language Research Unit, University of Birmingham.) Bazerman sees his work as leading to practical knowledge for writing teachers, and as he has produced these specialized sociological studies he has also produced an influential textbook for university writing courses, *The Informed Writer* (Boston: Houghton Mifflin, 1981; 1985; 1989).

sons, whereas we believe in neuroendocrinology because it is true. Many of these researchers critical of Merton trace their inspiration to Thomas Kuhn. But there have been a number of other strands entwined with this approach: the sociology of knowledge from Karl Mannheim, ethnographic methods from anthropologists studying nonwestern cultures, and conversation analysis from Harvey Sacks and others. Rather than trace all these strands I shall contrast two studies with Merton's.[14] Here I am not so much concerned with the specific findings of these studies as with their general method. They represent two ways of keeping scientific knowledge from seeming natural and self-evident: by focusing on controversies and contexts, and by taking apart the processes by which facts are constructed.

Steven Shapin's "Pump and Circumstance: Robert Boyle's Literary Technology" changes our view of a scientific fact by looking at its production in a political, social, and textual context. Shapin argues that Boyle and his colleagues, attempting to promote the experimental approach to natural science in a seventeenth-century England in which this approach was still controversial, designed their program to produce "indisputable matters of fact." They were produced through a material technology, such as the expensive and delicate air pump, a literary technology, in the form of reports that would make it seem as if the reader had witnessed a demonstration, and a social technology that governed debate such that matters of fact would be decisive.

So far Shapin seems similar to Merton, in his focus on the seventeenth century as crucial for modern science, his interest in the social validation of knowledge, and his interest in the rise of publication along with the rise of science. The key difference is that Shapin argues that we must look to the social and political struggles of the seventeenth century, not only for the origins of social institutions like the Royal Society, but also for the origins of matters of fact. He describes, not a unitary, functional science that progresses smoothly to the science of the present, but a variety of fundamentally different sciences competing with each other, varieties we see now through the assumptions of modern textbook science.

> The foundations of knowledge were not matters merely for philosophers' reflections; they had to be constructed and the propriety of their foundational status had to be argued. The difficulties that many historians evidently have in recognizing this work of construc-

14. See Steve Woolgar, *Science: The Very Idea* (London: Tavistock, 1988) for a brief summary and critique of various strands.

> tion arise form the very success of that work. To a very large extent, we live in the conventional world of knowledge production that Boyle and his colleagues amongst the experimental philosophers laboured to make safe, self-evident, and solid. (P. 482)

In Shapin's study (and in his larger-scale work with Simon Schaffer, *Leviathan and the Air-Pump*) the alternative science is that of Hobbes, who objected to the experimental method and held out for a more logical basis of scientific thought; in other studies, the other side of the controversy might be proponents of "charm" in high-energy physics, the geological catastrophists, or phrenologists, or J. Barkla, a twentieth-century physicist whose discovery of a "J phenomenon" was not accepted by the rest of the scientific community.[15] To follow a study like Shapin's, we have to suspend our certainty that Boyle won because he was, after all, right about the properties of air.

Each of Merton's norms of the ethos of science is shifted by Shapin, from the terms of a functioning system to terms of a contest between opposed forces. Merton's Universalism corresponds to Boyle's attempts to find a language in which his science—based on "matters of fact"—would gain support by being presented as a neutral ground between bitterly divided sects. Where Merton talks about Communism, in terms of channels for the sharing and evaluation of given information, Shapin talks about texts, focusing on the medium and its powers. So, for instance, he looks at Boyle's "prolixity" as a technique for creating a sense of realism, a sense that the reader is a "virtual witness." Merton considers Disinterestedness as a norm of the scientific community, whereas the parallel category in Shapin's analysis, "modesty," is another rhetorical device. For Merton, "Organized Skepticism" can be explained as another functional norm, whereas Shapin treats the apparent skepticism of Boyle's separation of the language of fact and the language of interpretation as a strategic device for the defense of his epistemology.

From Shapin's viewpoint, the norms of science tell us about its

15. Andrew Pickering, *Constructing Quarks: A Sociological History of Particle Physics* (Edinburgh: Edinburgh University Press, 1984); Steven Yearley, "Representing Geology: Textual Structures in the Pedagogical Presentation of Science," in *Expository Science: Forms and Functions of Popularization,* ed. Terry Shinn and Richard Whitley (Dordrecht: D. Reidel, 1985), pp. 79–101; Steven Shapin, "The Politics of Observation: Cerebral Anatomy and Social Interests in the Edinburgh Phrenology Disputes," and Brian Wynne, "Between Orthodoxy and Oblivion: The Normalization of Deviance in Science," both in *On the Margins of Science: The Social Construction of Rejected Knowledge, Sociological Review* Monograph 27, ed. Roy Wallis (Keele: University of Keele, 1979).

rhetoric, not about its ethos; to accept them as functional is to promote an ahistorical view of science and to accept uncritically science's view of itself. Merton's and Shapin's approaches may seem complementary, one starting at the level of the whole—institutions—and looking at the parts—the individuals—the other starting from the parts and working up to the whole. But for Shapin, the acceptance of science's view of itself keeps one from seeing how it works on any level—one not only has to untangle complexity, one has to penetrate disguises. "Just as the three technologies operate to create the illusion that matters of fact are not man-made, so the institutionalized and conventional status of the scientific discourse that Boyle helped to produce makes the illusion that scientists' speech about the natural world is simply a reflection of that reality" (p. 510). So the relation of Shapin to the texts he studies is radically different from that of Merton; Shapin wants to see more than seems to be there, to see through the text to another set of meanings. For Shapin, as for many analysts of culture, the text conceals its origins.

Even traditional historians who would not accept the "Strong Programme" have long been concerned with the political and social background of science. It is harder to see this background in contemporary science, where the world of scientific facts may seem divorced from the world of political and social struggle. However many meticulous historical case studies make the connection, it will always be possible to say that science is different now. The sociologist Harry Collins comments on this apparent lack of evidence for the social construction of contemporary science. "It may be that scientific institutions have become more autonomous, so that the social network between science and the wider society is now sparse. I think it far more likely that it is a matter of not being able to 'see the wood for the trees' in very recent scientific history" (p. 153). An influential article by Collins, "The Seven Sexes," shows how a study of a current controversy while it is still unresolved can lead to issues relevant to textual study (even though, as we shall see, Collins excludes written texts from his study).[16]

Collins describes the problem of what Shapin calls the "self-evidence" of scientific fact using the figure of a ship in a bottle:

> It is as though epistemologists were concerned with the characteristics of ships (knowledge) in bottles (validity) while living in a world where all ships are already in bottles with the glue dried and the

16. Harry Collins, "The Seven Sexes: A Study in the Sociology of a Phenomenon, or the Replication of Experiments in Physics," *Sociology* 9 (1975): 205–24.

> strings cut. A ship within a bottle is a natural object in this world, and because there is no way to reverse the process, it is not easy to accept that the ship was ever just a bundle of sticks. (P. 94)

Although Collins agrees with Shapin that the analyst needs to cut through these assumptions, they have quite different methods, in these studies, of taking apart the ship in the bottle. Shapin and other historians are able to connect scientific controversies to their social context by studying controversies sufficiently far in the past for their social context to assume definite outlines for us. The terms of the historical background—class, political struggle, educational and religious institutions—are dealt with in familiar social historical terms, and in that sense can be taken for granted. Collins, on the other hand, sees an advantage in focusing on contemporary controversies. Even though tracing of social interests in conventional sociological terms is not as easy in such cases, the analyst has the advantage that the process of making a self-evident fact is not yet complete. "It is actually possible to locate this process in scientific laboratories, in letters, conferences, and conversations. It is possible then to perform a kind of automatic phenomenological bracketing for ideas and facts, by looking at them while they are being formed, before they have become 'set' as part of anyone's (scientific) world" (p. 95).

Collins privileges the conversations, letters, and interview comments of scientists—what he calls the contingent forum—over the published articles—part of what he calls the constitutive forum—because the informal texts can show what goes on before discourse is fitted into the formalities of research articles. His style of interviewing is an important part of this strategy. Sociologists usually find out about the social side of science by asking about it directly; their interviews or questionnaires are designed to get reliable data about institutional and interpersonal issues. But when one focuses attention on it in this way one gets certain familiar and stereotypical answers. Collins approaches the social obliquely; his interviews "were built around detailed *technical* discussion of the experiment and scientists' interactions rather than straightforward sociometric questioning" (p. 98). So his approach is exactly the opposite of that of Mertonian interviews—he finds the social through the technical instead of extrapolating from the social to the function of norms and social structures in the technical realm.

Collins' study has implications for textual analysts even though it avoids written texts. He compiles lists of interview comments to show the varying sources of attitudes toward an experiment and the use of "other than formal methods of argument and persuasion" (p. 106).

These lists would lead us to look in the rhetoric of (informal, contingent) scientific discourse for the full range of rhetorical appeals used in other discourses, and not just for the logic and evidence on which scientific arguments are supposed to be based. Collins is interested in these rhetorical devices in relation to the problem of replication, arguing that what scientists are engaged in is "negotiations about the meaning of a competent experiment in the field" (p. 107). He sees a circular rhetoric in any field of science that is making fundamental discoveries: only a competent experiment will show the true nature of the phenomenon, but the competence of the experiment is judged on whether it shows the true nature of the phenomenon. Thus the argument could go on forever, but as Collins shows in a subsequent article, in practical terms closure is brought by social means.[17] Somebody doesn't get their articles cited, or their grant renewed, or their discovery in the textbooks.

Scientific Texts: Discourse Analysis and Its Critics

Shapin's and Collins' approaches seek to take us behind the text; in some ways this approach is much like literary study when it focuses on manuscripts, letters, and biographical background as showing the process underlying the published work. I need, for the questions I am asking, the example of someone who looks at the surface of the text. A key work in this shift of focus is Bruno Latour and Steve Woolgar's ethnographic study of biochemical research, *Laboratory Life*. Latour and Woolgar see the whole work of science in terms of the production of inscriptions: machines tracing graphs, researchers making notes, articles lying on a desk. Latour has continued this approach in his work on Pasteur and has outlined it for a general audience in *Science in Action*.[18]

17. Harry Collins, "Son of the Seven Sexes: The Social Destruction of a Physical Phenomenon," *Social Studies of Science* 11 (1981): 33–62.

18. Bruno Latour, "Give Me a Laboratory and I Will Raise the World," in *Science Observed: Perspectives on the Social Study of Science*, eds. Karin D. Knorr-Cetina and Michael Mulkay (London: Sage, 1983); *Les Microbes: Guerre et Paix* (Paris: Éditions A. Métailie, 1984).

A nonsociologist interested in sociological approaches to scientific texts might start with Bruno Latour, *Science in Action*; Gilbert and Mulkay, *Opening Pandora's Box*; Charles Bazerman, *Shaping Written Knowledge* (Madison: University of Wisconsin Press, 1988); Bruno Latour and Steve Woolgar, *Laboratory Life* (Beverly Hills and London: Sage, 1979); Michael Lynch, *Art and Artifact in Laboratory Science*; and K. Knorr-Cetina, *The Manufacture of Knowledge* (Oxford: Pergamon, 1981).

The break with both Shapin and Collins implied in this approach through texts can be seen in Woolgar's article, "Discovery: Logic and Sequence in a Scientific Text."[19] Woolgar's "central assumption" is the "isomorphism between presentational context and scientific concepts" (p. 239). That is, a text on a phenomenon takes on the same shape as that phenomenon, or rather, the phenomenon takes shape through the text. So the text is neither the empty tube that carries the scientific facts (as it is for Merton), nor is it the formal surface that conceals the real business of science (as it is for Collins); it is the crucial document that allows Woolgar to recover "the structure of the conceptual model which is made use of in recognizing that [a discovery] is what it [the phenomenon] is" (p. 251) . The search for the real processes behind the apparent processes of science will yield just another text, not a truer one. Woolgar, like some literary critics, insists that one cannot get to something beyond representation. But such an insistence may be more surprising to Woolgar's sociological readers than it would be to many literary critics. When he says in a note, "I have no interest in the 'accuracy' of the data," he is tossing out most of what sociologists do, but he is keeping the privileges of a literary critic. What he seems to mean is that he is not interested in whether the speaker's account of discovery accords with some hypothetical objective view of events, but is interested, rather, in how the events are defined, and are constantly redefined in further texts, as a discovery.

Woolgar makes it clear in his heavily ironic introduction that this approach contrasts both with that of "the rationalists," like Merton, and "the Strong Programmers," like Shapin. Woolgar notes that both sides, in their discussions of the nature of proof and evidence, refer to documents for their evidence: "To the extent that we are constrained in our use of available language resources, we will inevitably reproduce the language of realism" (p. 242). It is that rhetoric Woolgar seeks to analyze. He argues that scientists themselves, in controversies, use both rationalist and Strong Programme arguments. He steps back from the sociological controversy about scientific knowledge to study the textual strategies used in all controversies to create a sense of reality. "The analysts' task is not to resolve such disputes, but rather to develop an appreciation of their form and currency" (p. 243).

Woolgar presents his study as an extension of Collins's, but it can

19. Steve Woolgar, "Discovery: Logic and Sequence in a Scientific Text," in *The Social Process of Scientific Investigation*, Sociology of the Sciences, Volume 4, ed. K. Knorr, R. Krohn, and R. Whitley (Dordrecht: D. Reidel, 1980).

also be read as a reversal of the direction of interpretation. Collins speaks of negotiating the character of the phenomenon, whereas Woolgar notes that "the out-there-ness of a phenomenon is accomplished in establishing its properties" (p. 245). This seemingly innocent turn means that instead of using texts to show the socially contingent nature of scientific phenomena, as Collins does, Woolgar assumes the social contingency of research, including his own, and uses this assumption as a starting point in analysis of texts. Woolgar uses the term "documents" in its special ethnomethodological sense, meaning all sorts of outward and visible signs from which the interpreter must begin.[20] But the document he analyzes here is a document in the familiar sense of the word as well; he says he has "arbitrarily chosen" to write about writing, when he could just as well find practical reasoning in "the actions, conversations, seminar discussions, conference presentations, inscriptions, recordings, and writings of scientific work" (p. 245). But as we shall see in contrasting him with another ethnomethodological approach, his choice of published writing as the scene of such practical reasoning is significant; he challenges the implied hierarchy in which writing, and especially a published text, is secondary to informal speech and practical actions. His choice of a text, the Nobel address of an astrophysicist, would seem to be both secondary and unrepresentative; secondary because it comes long after the events it describes and their announcement, and unrepresentative because not many astrophysicists have to produce Nobel addresses. But his point is that the events have to be redefined as a discovery in each new text, so that a late text does work just as the first publication did. And if the specific occasion is an unusual one, it is not unusual for a scientist to have an occasion to present a scientific claim in terms of a narrative of his or her career.

Woolgar focuses on four regular patterns in the text that he relates to accomplishing the "out-there-ness" of the discovery. "Preliminary instructions," such as the title, the identification of the author, and the occasion, assert that the text is *about* something, and direct us to consider it in only a certain context. "Externalizing devices," including the quasi-passive voice and the invocation of community membership, seem to deemphasize the author's role in the discovery, so that the discovery is seen in terms of a path of coincidences. "Pathing devices" are ways of "portraying work as the latest in a long line of

20. The classic source for ethnomethodology is Harold Garfinkel, *Studies in Ethnomethodology* (Englewood Cliffs, N.J.: Prentice-Hall, 1967); see John Heritage, *Garfinkel and Ethnomethodology* (Cambridge: Polity Press, 1984), for an introduction.

development" (p. 256), shaping history retrospectively. And "sequencing devices" are forms that "act as a cutting out process, whereby other potential paths and potentially relevant events are backgrounded" (p. 258). Woolgar's conclusion is that such narrative devices that create a sense of "the nextness" of events produce the effect of logic in scientific texts, so as to make it difficult to imagine alternate interpretations. Woolgar, like a good literary critic, gives striking examples from the text to illustrate each of these devices. But for all his close reading, his devices are a linguistic grab bag, hard to define in terms of signals in the text. Instead, one has to start with the sociological problem, and then look for the bits of text that relate to it. This does not undermine his approach as sociology, but it makes it difficult to use in other disciplines, such as linguistics, which require that generalizations be tied to formal features in the text.

At the end of his essay, Woolgar suggests that "the perspective here might be profitably developed and extended to an examination of a much wider range of scientists' accounting practices" (p. 263). This was done in a series of articles by Nigel Gilbert and Michael Mulkay, who, like Collins and Woolgar, based their conclusions on a detailed study of one research program, in this case a biochemical controversy concerning the mechanism for transfers through cell membranes. Like Woolgar, they were interested in the ways scientists define their world in their accounts of it. Like Woolgar, they looked at patterns of the accounts themselves, instead of using these accounts directly as evidence for the correctness of one or another assertion about the social structure of science.[21]

Gilbert and Mulkay's major device is the categorization of scientists' discourse into an "empiricist repertoire" stressing impersonality and experimental results, and a "contingent repertoire," which acknowledges social factors. These are like the positions of Mertonians and of Strong Programmers as Woolgar lays them out at the beginning of his article. Gilbert and Mulkay argue, as Woolgar does, that scientists can shift strategically between these two repertoires. Thus the repertoires coexist in the same texts, instead of being found in separate places like Collins' contingent and constitutive forums. Gilbert and Mulkay show such shifts by contrasts of various texts. It is the comparisons, rather than any specific linguistic features, that make the devices apparent. For instance, in one chapter which I shall cite frequently, "Accounting for Error," they show how scientists

21. The essays are collected in Nigel Gilbert and Michael Mulkay, *Opening Pandora's Box*.

regularly use the empiricist repertoire to describe their own successes, while using the contingent repertoire to explain why competing researchers are wrong. In another chapter, they describe a rhetorical device that allows for contingency in research while still preserving the empiricism of science, the "Truth Will Out Device," holding that facts will triumph over social factors in the long run. They also analyze attributions of consensus, illustrations, and even jokes using these categories.

Though my own approach follows that of Gilbert and Mulkay in many ways, I now find their analysis into two repertoires a cumbersome analytical tool. It does not help in studying the generation or reception of the text. The problem that bothers me is not the problem that bothers more empiricist sociologists of science like Harry Collins, the turning of irony upon irony that characterizes this deconstructive approach.[22] Rather, my main problem is that the two categories seem to owe their existence to a polemic against the idea that anything lies beyond the text. Even if one is persuaded by this polemic, it does not necessarily take one much further in textual analysis. As Woolgar shows clearly, he and Gilbert and Mulkay turn a sociological controversy into an analytical tool. This leads them to striking insights but means they are tied to the terms of that controversy. Like Woolgar's contrast between "rationalist" and "Strong Programmer," the two repertoires seem to lead only to the selection of some features that parallel these two lines of interpretation. And Gilbert and Mulkay's interpretations seem to be limited to showing that there is interpretive variability; beyond making this sociological point, they say little about the processes of writing and reading.

There are two lines of criticism of discourse analysis in the sociology

22. For further discussions of the differences of approach between Discourse Analysis as practiced by Gilbert and Mulkay and their colleagues and the Empirical Program of Relativism as practiced by Harry Collins, see Michael Mulkay, Jonathan Potter, and Steven Yearley, "Why An Analysis of Discourse Is Needed," in Knorr-Cetina and Mulkay, eds., *Science Observed.* Collins' response, "An Empirical Relativist Programme in the Sociology of Scientific Knowledge," is in the same volume. Mulkay plays wittily on the debate in *The Word and the World* (London: Allen and Unwin, 1985), and Malcolm Ashmore critiques it in his playful essay, "The Life and Opinions of a Replication Claim: Reflexivity and Symmetry in the Sociology of Scientific Knowledge," in Steve Woolgar, ed., *Knowledge and Reflexivity: New Frontiers in the Sociology of Knowledge* (Beverly Hills and London: Sage, 1988), pp. 125–54. Another critique of Gilbert's and Mulkay's approach to science studies is the review by Peter Halfpenny of *Opening Pandora's Box* and *The Word and the World,* "Talking of Writing, Writing of Writing: Some Reflections on Gilbert and Mulkay's Discourse Analysis," *Social Studies of Science* 18 (1988): 169–82.

of scientific knowledge that I will keep in mind without really answering: one comes form the discourse analysts themselves, and another from ethnomethodology. Discourse analysts take an ironic stance toward scientific texts, but they also take an ironic stance toward texts in the sociology of science, toward all the devices of interpreters establishing an extratextual reality on which to base their arguments. And this irony applies to the sociological analyst's own text, including the present one; Woolgar has continued to investigate the implications of turning one's methodological tools back on oneself (see *Science: The Very Idea*), coming finally to a critique of the notion of the self. In these investigations he uses the ethnomethodological term "reflexivity," but literary critics unfamiliar with sociological jargon will still recognize the move. The consistent development of the discourse analyst's ironic stance is in Mulkay's work *The Word and the World*, in which he plays with his own text's attempts to establish authority by writing parodies and dialogues. For instance, "The Scientist Talks Back" is a one-act play that draws on the writings of Zuckerman, Collins, and Karin Knorr-Cetina, along with comments of scientists made in interviews, to present a fictional dialogue about replication in science and in sociology.[23]

Just as sociology of science has its deconstructivists like Woolgar and Mulkay, it has its phenomenologists—the ethnomethodologists. Michael Lynch, who was trained in ethnomethodology, has produced a series of studies that question all the lines of work that I have surveyed so far. One article, "Technical Work and Critical Inquiry" (1982), summarizes his critique and provides a model of an alternative approach.[24] Essentially, Lynch wants to study the way scientists themselves make sense of their world in their daily work. So he dismisses at the outset those sociological studies like Merton's that do not involve "the technical work of science." But he sees a flaw in the approaches that try to show the social contingency of scientific knowledge, arguing that they just impose sociological categories on scientific work. Like Woolgar, he says that the Mertonians and the Strong Programmers ignore the social analysis the scientists themselves do in their daily work. They assume some place to stand ironically outside science to see through its pretensions. But what validates the methods of sociology?

23. Woolgar collects some essays developing, exemplifying, and criticizing reflexivity and new forms in *Knowledge and Reflexivity*; see especially the contributions by Malcolm Ashmore, Anna Wynne, and Bruno Latour.

24. Michael Lynch, "Technical Work and Critical Inquiry: Investigations in a Scientific Laboratory," *Social Studies of Science* 12 (1982): 499–533.

Lynch proceeds through a critique of the various devices used to reveal social contingency, presenting all these devices as ways of getting around the fact that sociologists are not themselves doing technical work. "Critical historiography," like Shapin's, substitutes "disengaged overview" for "technical access." "In this way the 'particulars' of the technician's work are disregarded, except in so far as they supply the documentary materials for an historian's operation of showing how the reputedly objective concerns of the technician are the legacy of a contentious (and in some ways capricious) social history" (p. 505). When the historian's perspective is substituted for that of the technician, the self-evident quality of scientific objects is broken open, but in the process everything gets lost except what can be used to show social contingency.

Lynch's treatment of the sort of rhetorical analysis I am doing in this book is brief but damning.

> On setting up these operations, the identification of science within the traditional distinctions between logic and rhetoric, truth and fiction, and fact and construction is *inverted* for methodological purposes. . . . Rhetorical analysis of the ostensively 'non-rhetorical' relies upon an argument which places 'objectivity' in quotes (or brackets) and subsumes [it] to the omnirelevance of rhetoric. (p. 505)

Again, the critical analyst is accused of privileging his own categories, here simply the ironic reversal of just those categories put forward by scientists. Lynch asks analysts of the rhetoric of science, "Rhetoric as opposed to what?" (p. 527n). The claim that scientists are using rhetoric is only interesting as an ironic debunking of the assumption that their discourse is especially "objective." Once one grants that this objectivity is something they create in their work, the claim that everything is rhetoric has little meaning.

Finally Lynch criticizes the device of "the stranger" used by Latour and Woolgar, and by many studies (like mine) that must view science from the outside to find a chink in the armor of its apparent self-evidence. Lynch questions the sort of reflexive turn that reduces all of science to inscriptions and then sees scientists as doing the same sort of literary interpretive work as the anthropological observer. This claim for the sufficiency of nonscientific methods, he says, is disproved by the Latour and Woolgar "stranger's" in incompetence as a lab technician. There is, then, some crucial technical skill that the sociological observer does not have.

In Lynch's view, all these analysts of science are guilty of trying to impose their own rather simple sociological categories on the complex processes of scientific work. Instead of imposing social science methods on natural science inquiry, they should become aware of the processes through which scientists create their world. "Scientists speak and act in each others' competent presence in ways that exhibit an untold-of richness and specificity to questions on the constructive horizons of specific objects under investigation" (p. 511). But in order to get at this richness, one must try to become competent at what the scientists do, and one must be at the same place and the same time as the scientific work being done, in the presence of the objects under discussion; in ethnomethodological terms, the inquiry must be situated. That is, one must enter into the world of the lab, and recognize one's own limitations as a nonscientist.

> To point to this disciplinary limitation on what can count as a competent observational vantage is not to propose closure on the problem of multiple interpretations, but instead to insist that any interpretation, evaluation, or argument must, first of all, contend with what scientific practitioners produce and recognize as a competent interpretation, evaluation, or argument in *their* local setting of inquiry. (p. 54)

Lynch gives an example of how scientists negotiate the character of an object.[25] In this case, a lab director and a technician are discussing an electron micrograph montage and determining whether it was done incorrectly and whether it is still usable. For textual analysts, the key point is that this discussion only makes sense when considered in the presence of the montage. The processes going on cannot be reconstructed later, just from written texts. In Lynch's view, such a situated analysis avoids the imposition of sociological frameworks but still shows the socially negotiated nature of reality. "The work, then, not only consists of a 'progress' through a socio-historical terrain; its *progression* from within involves an articulation, for all practical purposes, of what that socio-historical terrain consists of as immediately pertinent details" (p. 519).

The implication is that a study that starts with a ready map of the social landscape—in terms of class interests, or theories of paradigm shifts, or progressive research programs, or the unmasking of

25. The example is discussed within a more detailed theoretical framework in *Art and Artifact.*

rhetoric—does not allow the work to articulate its own map. Like Merton and Woolgar, Lynch sees science as social, but he sees the relevant items of the social framework as emerging in the situated work of scientists, rather than in the work of sociologists or historians fitting science into their categories. He sees ethnomethodological "documents" as the basis of analysis, but criticizes approaches that depend entirely on written and published accounts, which cannot yield analyzable material on scientific practices in action.

Lynch also differs from Woolgar in the relation he tells the interpreter to take toward the text. In the view of the ethnomethodologists, the interpreter must drop the ironic stance and enter into the technical work of science undefended to see how natural objects are interpreted. This critique of irony is the most valuable part of Lynch's approach for me, for I find that, far from unmasking the ideological processes of science, I have created descriptions that don't surprise the biologists I am studying. Surely if I were unmasking some concealed pattern, they would be annoyed, or would at least disagree with what I showed. While I would have preferred to reveal what they could never have seen, I have to admit that they are quite capable of themselves undertaking the sort of analysis that I do.

I have not taken the ethnomethodological program as the basis for this study for three reasons. First, I can't. The ethnomethodologists, in accordance with the severe demands of their methodology, set very high standards for initiation into their fraternity, so that Lynch, for instance, must try to be both a phenomenologist and an electron microscopist. To an outsider, at least, these standards seem not only high, but unattainable. Second, I'm not sure that even if I could take this approach, I would find anything interesting. Lynch's studies so far suggest what might be done, but the findings themselves are not so interesting as the methodological preamble. Finally, I remain suspicious of ethnomethodologists for the same reason that many literary critics are suspicious of phenomenological criticism; it seems to posit an underlying level of reality in processes, a level that seems neither reachable nor necessary. In this case, one could ask whether Lynch can escape imposing social science methods like all the other approaches he criticizes. Lynch, I must say, argues strongly and consistently against any accusation that he claims to find a deeper level of reality. Still, one sometimes thinks when reading ethnomethodologists of the Emperor's new clothes, or of a new version in which the Emperor claims to walk in front of his people completely naked, but in fact is wearing rich old robes.

I have arranged this sequence of studies so that it leads back to my own approach. Like Bazerman, I study scientific texts as part of a social system. I am, like the Strong Programmers, involved in a project of explaining the scientific in terms of the social. Like Collins, I see this as requiring a relativist epistemology, a disorienting shift of perspective, and detailed study of technical work. Like Woolgar, Gilbert, and Mulkay, I see texts as a way of investigating these social negotiations, and I would study, not just the content of texts, but their form and processes of production. But like Lynch, I want to avoid the ironic stance of these other approaches. The project should also require some attempt to relate the conflicts involved in science to larger conflicts in society, whether the relation is one of simple congruence or is more complicated. In fact, I make these connections only occasionally; I shall return to these relations in the Conclusion.

Materials and Methods

Each of the studies in this book uses the form of some scientific texts to reveal the social processes of the construction of scientific knowledge. Though the questions I am answering and the theoretical basis for my approach are drawn from sociology of scientific knowledge, I find that my methods remain those of someone trained in literary criticism and rhetoric. Like many traditional literary historians, I am particularly interested in revisions, and in sources and analogues, as starting points for analysis. In various studies I draw on such traditional topics of literary and rhetorical study as ethos and pathos, narrative, irony, persona and characterization, subjects and tenses of sentences, and the use of quotation and echoic speech. In effect, I take each text apart, so that we can see alternative choices at various points, and then put it back together again in a way that stresses a coherent pattern in these choices, and implies a process underlying them.

The studies that follow analyze writing in various genres by five American and British biologists whose research touches in one way or another on evolutionary questions. I chose to focus on evolutionary biology simply because that was what my first two subjects had in common. But biology is a useful discipline for a study of this kind. First, it is not physics, and although many sociological and philosophical studies take physics as the exemplar of science, I think we may learn from the rather different methods of biology. Evolutionary biology is often nonexperimental, so it raises rather different questions about observation, interpretation, and persuasion than those treated in most studies of physics. The processes these researchers want to explain are entirely inaccessible by direct means (one researcher points to cosmology as a parallel example), so there can be no naive sense for them of immediate confrontation with nature. Narrative is the point of the discipline, and the need for interpretation to make the mass of data coherent is acknowledged by the scientists themselves. I will refer in the chapter on popularizations to Ernst Mayr's remark in *The Growth of Biological Thought* that advances in biology involve conceptual shifts rather than discoveries; the same facts come to be seen from a radically different perspective. This makes rhetorical analysis of the work of evolutionary biologists particularly interesting, since the finding can be said to emerge in the text and in discussion, as events are assembled into a narrative, rather than, say, having the finding seem to emerge, its meaning self-evident, from the detector of an accelerator. Finally, evolutionary biology has long been an area of public interest and controversy, so its larger ideological context is of interest. For instance, the popularization of evolutionary ideas in sociobiology has an effect on a public debate, on social and political issues, in a way that popularization of the equally interesting issues of quantum physics does not.

In the course of this study I have talked to a number of biologists about their writing, but I have chosen to discuss only a few in this book. One thing they all have in common is that they work on some boundary between fields; this creates a tension that brings out some of the social networks involved in scientific publication. I shall introduce them all here to give a Dramatis Personae for the studies that follow:

David Bloch is an example of an experienced researcher trying to enter a new area of research relatively late in his career; his writing showed all the skills of an experienced member of the discipline, but he lacked, at least in the beginning, the network of contacts that even a newly graduated Ph.D. would have. His original field was cell biology; he studied at Wisconsin and Columbia and taught at UCLA before coming to the Botany Department of the University of Texas in 1961. Until 1980, his published articles were on cell biology, and through the 1970s his lab was supported by grants for flow cytometric studies. He also did a relatively large amount of undergraduate teaching. The origins of the "big idea" that led him to change his field of research are complex. He traced his interest in evolutionary questions to his reading of Schrodinger's *What is Life?* in graduate school, to his having to present basic concepts in introductory biology courses, to a period of free writing forced upon him by a back injury that kept him out of his lab ("I was prone to write," he said), and to a graduate seminar that allowed him to follow up some of the questions raised during this free period. His ideas on the evolution of nucleic acids emerged before he had a chance to read the current literature, instead of arising in response to that literature; this may explain why it was so hard for him to fit these ideas back into the current questions in the field. As we shall see, he wrote a number of drafts of an article, and made a number of proposals for new funding, while "moonlighting," as he put it, with his lab funded for cell biology work. His articles began to attract some interest in major journals and a news report in *Science* in 1985, after he began a collaboration with a physicist—surely a rather unexpected linking of disciplines—that enabled them to present more elaborate statistical analyses of the data. He got some funding from a private foundation for his new line of work, and prepared several further articles. Dr. Bloch died of cancer in October 1986, after a year of treatment during which he continued to do a great deal of writing and research.

David Crews specializes in the reproductive physiology of reptiles, so he sees himself as working between the herpetologists who study snakes and lizards from a natural history standpoint, and the neuroendocrinologists who compare various hormonal control systems. He began his graduate training in the Institute of Animal Behavior at Rutgers, after studying to be a social worker. It was there he began his physiological study of reptiles, working under the comparative psychologist Daniel Lehrman (whose name will come up later in the controversy over sociobiology). His Ph.D. was in psychobiology, an interdisciplinary area of research combining physiology, comparative psychology, and ethology. He did postdoctoral studies at Berkeley and taught for seven years at Harvard before coming in 1981 to the University of Texas, where he is now a Professor of Zoology and Psychology. As we shall see, this need to look in two directions for his audience complicates his publication of some of his articles. He directs a laboratory with several postdoctoral students and a number of graduate and undergraduate students. He has several research grants, which is useful for my purposes because he must devote much of his writing to getting his funding renewed or finding new sources. He is author or coauthor of about ten or fifteen papers a year. Some of these are popularizations and reviews aimed at presenting his area of work to a larger audience; he remarks that his applications for funding have made him aware of the need to be able to explain and justify the goals of his research simply.

Lawrence Gilbert specializes in the population biology of tropical forest insects and plants, a line of research that cuts across several of the older fields of zoology and botany. After receiving his undergraduate degree at the University of Texas in 1966, he had a Fulbright scholarship to Oxford, so that he has some links with research traditions in England, which in this area are not identical to those in the United States. He began publishing articles while he was a graduate student at Stanford in the late 1960s. He is now a Professor of Zoology at the University of Texas, where he also directs the Brackenridge Field Laboratory, a reserve in Austin. The research on which I will focus concerns the behavior of *Heliconius* butterflies and the morphology of the passion vines on which they lay their eggs; he has studied the relations of these butterflies and vines both in the jungles of Costa Rica and in a laboratory setting. As a population biologist, he is interested not just in observing individuals or describing a species, but in showing how and why populations vary in their natural setting. His work presents a particularly interesting case in popularization—a colorful subject matter combined with a highly abstract theoretical problem.

Geoffrey Parker's career also straddles a boundary. Like Dr. Bloch, he was moving into another field, teaching himself its methods and finding other researchers who shared his interests. His training, at the University of Bristol, was in entomology—he did field studies of the mating habits of dung flies. But these studies led him to broader problems of applying mathematical models, drawn from game theory, to the evolution of many sorts of behavior, such as fighting strategy, the competition of sperm, and the competition of parents and offspring. He commented to me: "The development of the new field involving game theoretical analysis of evolutionary problems is generally accepted as starting with [John] Maynard Smith and Price in 1973, though several people had made contributions in this direction beforehand (e.g., in investigating evolutionary problems of a frequency-dependent, game-like nature). I started modelling dung-fly mating and other more general problems before the formal ESS [Evolutionarily Stable Strategy] approach was available, and had to 're-jig' in terms of the new formalism." Dr. Parker's work has been highly influential in this developing field, and his articles are frequently cited in the sociobiological literature. He had one year of sabbatical working with the King's College Sociobiology Research Group in Cambridge, but has otherwise spent his career working at the University of Liverpool, where he is now Reader in Zoology.

The study of E. O. Wilson's *Sociobiology* took a different form from the others, because I interviewed neither Wilson nor any of his critics and drew the interpretations of the text only from published comments. (I did get his comments, and those of some critics, after I had written the chapter.) Fortunately, there is a good biographical account by Ullica Segerstrale, "Colleagues in Conflict," that clarifies some of the more baffling strands in his career.[26] Segerstrale stresses the shaping influences of the Society of Fellows at Harvard, where Wilson was inspired by the attempts of the entomologist W. M. Wheeler "to integrate the social and natural sciences on the basis of equilibrium theory." Along with this "scientific agenda," Segerstrale attributes to Wilson a "cognitive approach" linking scientific and moral notions, a "personal moral agenda" deriving from his upbringing as a Southern Baptist in Alabama and his "reconversion" to evolutionary thought (p. 56–57). His attempt to produce a grand synthesis in *Sociobiology* was so unusual that another sociobiologist, Robin Dunbar, suggested I study it, considering his rhetoric in its disciplinary context.

Though a British zoologist recommended the study of *Sociobiology* to me, Wilson's line of evolutionary thought differs in important ways from that of Parker and other British sociobiologists, just as it differs in important ways from the line in which Crews and other American comparative psychologists place themselves, and from the population biology of Gilbert. By lumping all these writers together as "evolutionary biologists" I do not mean to suggest that the term describes one unified program or self-defined discipline. What these researchers do share are the problems of bringing together several disciplinary perspectives, and certain basic assumptions about what sorts of evidence and forms are persuasive.

My selection of subjects and my selection of textual features could both be criticized as idiosyncratic by researchers who want more generalized knowledge about society or about texts. First there are the criticisms of traditional social scientists who identify knowledge with quantitative methods. An administrator who commented on an early research proposal, based on the work in my first two chapters, said, "What do you expect me to make of a study with an *n* of 2?" Perhaps my indifference to this problem is the result of my literary training. Like some of the other analysts of texts work I have described, I have based my findings on a very small sample, nothing like the thousands Merton considered necessary as a basis for conclusions. But with such case studies, what number would be large enough? These four biologists (five counting Wilson) are different enough from each other to make for some interesting comparisons. They all write a lot and are successful in their specialized fields. But they are not chosen to be representative. Rather, I turned to each of them when I first heard

26. U. Segerstrale, "Colleagues in Conflict: An 'In Vivo' Analysis of the Sociobiology Controversy," *Biology and Philosophy* 1 (1986): 53–87.

about some situation or problem he had as a writer. And they all agreed to work with me, so they must all have had some unusual interest in finding out about the ways in which scientists write. I will have to take refuge in the argument David Crews makes for his studies of atypical species of reptiles—it is the unusual cases that are likely to lead us to form hypotheses that will challenge accepted ideas. These hypotheses can then be applied to the more typical species.

The texts chosen are also, in some ways, atypical. For instance, most articles were not rejected five times, like those I study, and most proposals are not revised over such long periods, and most popular articles are not so extensively rewritten by the editors. But I shall try to show that in these rare, open conflicts one sees social processes that are also at work in other texts, but which are less easily seen when all goes smoothly. When texts seem to emerge unproblematically from the research work, the ways in which they emerge, the choices that have to be made, are not apparent, even to the writers and readers themselves.

I have chosen to study texts from several different genres—experimental reports, review articles, proposals, popularizations, and one massive and unclassifiable book. Most recent studies have focused on the experimental article (most importantly Bazerman's *Shaping Written Knowledge*). It makes sense if one is trying to demonstrate the importance of rhetoric in scientific writing to start with what is apparently the most scientific and least rhetorical form. It should not surprise anyone to hear that grant proposals are rhetorical, or that popularizations require careful consideration of audience, but some readers, particularly nonscientists, might think that research reports just communicate facts. Still, I think that the rhetorical nature of such articles should be well established by the studies I cite, and I can go on to other genres. I am more concerned with the relations between texts in several genres, with the production of knowledge as a social process that includes (in Ludwik Fleck's terminology) both the esoteric audience, the core group who will read the original report, and the exoteric, the broader community concerned with science who will read reviews, popularizations, proposals, textbooks, and news reports.

There is another line of criticism of my methodology that brings out an important feature of my approach. Linguists may find that when I analyze passages I ignore all that really interests them in the nominalizations, passive or active constructions, hedges, cohesive devices, and shifts of verb tenses. For each feature I interpret there are of course many possibly interesting features about which I say nothing at all. What brings a feature to my attention is the difference between a text and its revised version, or between a text and a comment on it, or between two texts for different audiences. My assumption is that such comparisons will bring out features that matter to the participants, rather than just features that are expected by the analyst. I do not provide a system that covers all the data, with unambiguous features in the data relating to each of my categories.[27] But I have set out, not to provide a general description of scientists' discourse, but to find textual evidence for the social nature of that discourse. If my approach is useful to linguistic discourse analysis, it is not in providing any formal system, but in suggesting how certain features might figure in social negotiation.

On such a methodological basis my claims must be tentative. I would hesitate to generalize any of my more specific claims about the links between linguistic features and social processes to scientific discourse as a whole; there are too many ways in which other cases may differ. On the other hand, I have become cautious about dismissing any of my observations as peculiar to one case. This is because again and again the scientists who have read my studies have said that what I say is true enough, but that it is true only of Americans, or of the British, or of the new young researchers, or of well-established, eminent researchers, or of evolutionary biology, which is taken as less scientific than physiology or molecular biology, or of biology in general, which is taken as less scientific than physics, or of those researchers who work with this genus of lizards. These comments may all be true, but I have now come across them in so many forms that I wonder if this process of categorization is a way of protecting the core of science from the suggestion of social contingency. Is real science, the unmediated encounter of man and nature, always going to be somewhere else, in another discipline, another age, another country? One advantage of starting with cases rather than with norms is that it is not my job to look for this somewhere else, for the typical science and the typical scientific text. I can begin with the material at hand.

27. For such criteria, see John Sinclair and Malcolm Coulthard, *Towards an Analysis of Discourse: The English Used by Teachers and Pupils* (London: Oxford University Press, 1974).

Chapter Two

Social Construction in Two Biologists' Proposals

Why begin a study of scientific writing with grant proposals? Research articles (which I consider in chapter 3) are usually taken as the central genre of scientific writing, and popularizations (chapter 5) are more familiar to nonscientists; proposals may seem to be purely administrative documents, necessary to the scientists and agency officials but unexciting to the sociologist of science. But for many scientists heading large laboratories, proposals are in one practical sense the most basic form of scientific writing: the researchers must get money in the first place if they are to publish articles and popularizations, participate in controversies, and be of interest to journalists. For these researchers proposal writing is by no means an occasional administrative duty; it is constant effort that may involve approaches to a number of different agencies, that may take about a quarter of the director's working time, and that requires more and more attention as grants are given for shorter periods, and fewer projects are funded.

Proposals are a promising place to begin a study of scientific texts in that they are the most obviously rhetorical genre of scientific writing: both writers and readers know that every textual feature of a proposal must be intended to persuade the granting agency. The rhetoric can be finely calculated because proposals are written for a very small audience. Most academic scientific research in the United States is funded by the National Institutes of Health (NIH) and the National Science Foundation (NSF), through a procedure of peer review in which researchers' written proposals are evaluated by panels of researchers from the same general field. The proposal is likely to be read only by the members of this panel, and by the Executive Secretary who administers the section covering the proposal topic.

Nonscientists have often raised questions about the fairness of such a system. John B. Conlan, a former congressman from Arizona, has charged, "It is an incestuous buddy system that frequently stifles new ideas and scientific breakthroughs," and Michael Kenward quotes an observer who says it is like a murder trial with "a jury of axe murderers from the same gang." There has been a great deal of study of the peer review procedures of these agencies, both by social scientists and by in-house committees.[1] Both the public criticisms and the studies have focused on what happens in the funding after the proposals are submitted, and have asked what role factors beside "quality of science" play in the decision to fund or not to fund a project; they ask, for instance, if there is an "old boy network," or if the panels tend to reject "high-risk" proposals. But the funding process starts before the panel even sees the proposal. I will argue that the writing of proposals, which takes up such a large proportion of the active researcher's time, is part of the consensus building essential to the development of scientific knowledge. To take the metaphor of the critic Kenward quotes, it brings the axe-murderers into the gang.

There is a paradox in the rhetorical strategy of the proposal, because the proposal format, with its standard questions about background and goals and budget, and the style, with its passives and impersonality, do not allow for most types of rhetorical appeals; one must persuade without seeming to persuade. And yet almost every sentence is charged with rhetorical significance. In classical rhetorical terms, the forms of appeal in the proposal are ethical and pathetic as well as logical; one shows that one is able to do the work, and that the work is potentially interesting to one's audience of other researchers, as well as showing that one is right. The writer describes the work so as to create a persona (a presentation of the author in the text) and insert the work into the existing body of literature. One has a special problem if one sees one's work as new or falling between two specialized fields; one must either present a persona as an established member of one of the fields, or redraw the fields around the work. In either case one places the potentially dissenting idea within a new consensus. The process of writing a proposal is largely a process of

1. See, for instance, S. Cole, R. Rubin, and J. Cole, "Peer Review and the Support of Science," *Scientific American* 237, no. 4 (1977): 34–41 (Conlan quotation, p. 34); J. Cole and S. Cole, "Which Researcher Will Get the Grant?" *Nature* 279 (1979): 575–76; S. Cole, J. Cole, and G. Simon, "Chance and Consensus in Peer Review," *Science* 214 (1981): 881–86; M. Kenward, "Peer Review and the Axe Murderers," *New Scientist* 31 May (1984): 13. F. van den Beemt and C. Le Pair, "Appraisal of Peer Review" (unpubl. paper) studied the mechanisms in the Netherlands. Joop Schopman, of the University of Utrecht, has written a detailed study of the proposals that led to the Center for the Study of Language and Information at Stanford: *The Foundation of the Center for the Study of Language and Information* (Utrecht: Rijksuniversiteit Utrecht, Department of the Epistemology and Philosophy of Science, 1988).

presenting—or creating—in a text one's role in the scientific community. Thus the texts of proposals may have something to tell us about how science changes and defines itself, as well as about how it is funded and how it is communicated.

I have collected all major drafts of proposals by two biologists at the University of Texas, David Bloch and David Crews (I shall analyze two of their articles in chapter 3). In one case, these were successive proposals submitted to several agencies over the course of eighteen months; in the other they were drafts of one proposal written and rewritten over the course of the ten months before its submission. In both cases the authors had comments of readers from which to revise, one using the peer reviews of previous attempts, the other getting comments from coworkers, so I collected these comments and the writers' responses to the comments. Thus I worked with three readings of the proposals: the writers', the readers', and my own.

For each proposal, I noted changes between drafts (first to second, second to third, etc.), sometimes following handwritten changes on the drafts. I categorized changes by what seemed to motivate them and noted especially those changes that seemed to indicate the writer's self-presentation or relation to the research community. I also noted changes of the content of the proposals, but as we shall see, there were few such changes (this might not be the case with many other proposals). In addition to these revisions, the authors made many of the improvements in readability any good writer might make. The changes affecting persona or context in the community are largely specific to the field, and as a nonspecialist, I had to have some clue from the writers or commentators to interpret them. And these categories are themselves matters of interpretation; my categorization of revisions represents one view of the text, which changed as I read on and as I tested my reading against other readings. I interviewed the writers about my interpretation of selected revisions, and I have also had them check, at various stages of my writing, the views in the study as a whole.[2]

2. I noticed, for instance, that in my earlier analyses, I tended to interpret almost all revisions as improving readability or accuracy, whereas later I tended to see more revisions as related to the author's self-presentation or place in the community. This shift may reflect a real difference between earlier and later drafts, as I show in discussing the authors' changes in strategy. It may also reflect a change in my reading in the course of the study. I began as a technical writing teacher, especially aware of ease of reading and precision of statements. As I read more drafts, comments, and letters, and especially as I interviewed the writers, I became more aware of the context in which these changes were made.

As the brief résumés at the end of chapter 1 show, the researchers I studied are in some respects representative of biologists at large research universities: both have supervised laboratories and have published many articles, and both have in the past received grants and reviewed grant applications themselves. But, as we shall see from the responses of referees to their articles (see chapter 3), their work is potentially controversial. The fact that they anticipated resistance to their proposals may have made them more self-conscious about their writing processes, and it certainly makes the rhetorical features of their proposals more apparent to a nonbiologist discourse analyst.

In chapter 3 we shall follow one of David Bloch's first publications in molecular evolution, a new area of research for him. His funding during this period was from a National Institute of General Medical Sciences (NIGMS) grant and from the NSF, for his flow cytometric studies. When he first set out to test his model, he used published data and a computer program written by one of his graduate students and did not have any outside funding for the work except for university grants for computer time and a semester off from teaching. As these studies developed, he applied to the NSF twice (versions 1 and 2 of the proposal) and to the National Aeronautics and Space Agency (NASA) for support, not getting funded but apparently getting much closer. Then he applied to the Public Health Service (version 3 of the proposal) and reapplied to the NSF.

Bloch had, in his favor, a successful laboratory and an original idea on a topic of great theoretical interest. He also had a fresh familiarity with the literature and a demonstrated expertise with, and access to, computers. All this impressed his more favorable reviewers. But as we have seen, he had not gone through a conventional apprenticeship in nucleic acid research; he was not known to the leaders in the field, and he was not oriented toward the structure and function studies that occupy most researchers in nucleic acid sequencing. The most critical of his early reviewers bluntly rejected his proposal as that of a newcomer to a field already full of people doing structure-function research. Until the *Journal of Molecular Evolution* published his article in December 1983, his papers in this area could only be listed as "submitted." And he had a clear enthusiasm for one model, which left him open to charges that he was jumping to conclusions prematurely, on the basis of insufficient data. Bloch was well aware of his rhetorical strengths and vulnerabilities from the responses he had gotten to his papers at conferences, to the articles he had submitted for publication, and to his proposals to various agencies. Like Crews, he was acutely conscious of increased competition for research funds. Bloch's proposals have the same sections as Crews's but since he did not have a long series of separate experiments to describe and justify, they are much shorter, following PHS page limits, with about ten pages of text. His collaborators in his group contributed comments and criticism, but he was entirely responsible for the writing. At the time of this study he had been working on the project for more than two years, with countless drafts (on a text processor) of four article manuscripts (see chapter 3) and five submitted proposals, in addition to applications to the university, between January 1982 and June 1983. Later versions of the proposals did not go through so many drafts. Though he accumulated more and more data in his sequence searches and responded or adapted to criticism, the proposals did not grow any longer. As we shall see, increasingly detailed discussions of data that he had gathered were substituted for passages that described the model.

David Crews's lab, with two postdoctoral and several graduate students, and half a floor of a building, is now funded by grants from NSF (for one species), NIH (National Institute for Child Health and Human Development) (for two other species) and a National Institute of Mental Health (NIMH) Research Scientist Development Award (for him). I studied his work between September 1982 and July 1983 on a proposal for a competitive renewal, after five years, of his current NIH grant. Although he had a strong record as a researcher, he was concerned about the scarcity of research funds. His panel would not be the same one that awarded him the earlier grant, and his work would not be the same work. He had received both enthusiastic and sharply critical responses to his current research reports, and he could not afford even one critical review of the proposal. He would have to face or to sidestep this resistance. Also, the increased competition for federal funds meant he needed to prepare for close scrutiny according to the interests of NIH. He would have to show that his work on lizards had fairly direct application to problems of human reproduction. He would have to justify his field work on behavior, a type of work for which he thought NIH had little enthusiasm, and he would have to show that the large number of experiments he proposed all made a single coherent research project. Since heavily funded researchers with several grants were getting increased scrutiny, he would have to justify the funding of his lab by both NSF and NIH, clearly separating his work on one species from his work on the others.

Crews's proposal was necessarily a long one, with more than ninety pages of texts and detailed lists of experiments and procedures. He took about four months, spending two nights a week, to write the first draft that he would circulate to his research group. He began with his earlier successful proposal and reviewers' comments on it, the NIH guidelines, a list of topics he wanted to include, some boilerplate (the technical writer's term for material that can be reused on many proposals) on materials and methods, and several of his recent review articles. For the main part of his proposal, the prospectus of experiments, he drew on summaries each of his assistants had written, describing their current work and plans. A letter he had recently written in support of his career award contained arguments on the benefits of his work to humans; this proved helpful in drafting the section on "Significance." After about two dozen drafts (done using a text-processing program from his handwritten revisions), he gave a version to his research group. He included the guidelines and reviews of his earlier proposal, since he considered the proposal-writing process part of the education of the postdoctoral and graduate students working with him. He explained the competitive situation: "This has got to be an orgasmic experience for a reproductive biologist."

The Writer's Persona and the Literature of the Discipline

The instructions for writing the body of the proposal included with the NIH application emphasize the panel's concern with ethos and pathos—the character of the writer as researcher, and the interest of his or her work to other researchers:

> Organize sections A–D of the research plan to answer the following questions. (A) What do you intend to do? (B) Why is the work important? (C) What has already been done? (D) How are you going to do the work? (Application)

Crews and Bloch are especially concerned with questions (B) and (C), defining their personae as researchers and the relation of their planned work to the literature. Both these criteria involve contradictions. The form of scientific reports, the syntax of scientific prose, and the persona of the scientific researcher all work against self-assertion. And the definition of scientific importance requires both that the work be original and that it be closely related to the concerns and methods of current research. We shall see these contradictions presented and resolved in the course of the authors' revisions.

The number of revisions each writer made is remarkable considering that the first draft I studied, in each case, was itself the result of many drafts. The number of drafts means little, when the writers are using word processors, but in the five versions of Crews's proposal and the four of Bloch's that I studied, there averaged five to ten large or small changes on each page of each draft, and hardly a sentence remained unchanged over the course of revision. I interpret these changes as improving the readability, defining the relation to the discipline, and modifying the persona.

One kind of revision we might expect to see, besides these three kinds of revisions, is substantive changes in the research proposed. But neither of these authors significantly altered his plan of work to counter possible criticism. Other NIH applicants do add to, delete from, or modify their methods sections, especially if they have gotten detailed criticisms on their pink sheets (pink sheets are the summaries of the study section's evaluation of the proposal, sent out after the decision). That Bloch and Crews did not revise their methods may indicate their relations to the specialties of the study section members, relations in which they are quite different from each other. Bloch's research is so unusual and so isolated from the mainstream that he got little detailed criticism. Reviewers suggested some statistical tests, which he then used, but their own work in microbial genetics seemed to give them little specific to say to help Bloch with his broad evolutionary questions. Crews, on the other hand, had been working for five years on the project for which he was requesting continuing funding, and had fifty or so pages of detailed descriptions of experimental work in progress; his specific methods had already proven themselves to the study group members, and if questions were to be raised, they would probably be questions of logistics and management rather than experimental design.

Many of the changes for readability would have been suggested by any editor. Both writers had served on grant panels, and had learned from the experience of reading piles of proposals that, in Bloch's words, they have to "get the idea across efficiently," and in Crews's words, "they have to be made exciting." For instance, Crews wants his sentences to "flow," so he deletes such unnecessary words as *causal agents* in the phrase, "Disorders of human sexuality are causal agents responsible for. . . ."[3] Both authors cut jargon wherever they

3. Often the biologists gave a different interpretation of a change than I did. This shows, not that they were right and I was wrong, but that the interpretation depends on the point of view, knowledge, and purpose of the reader, as well as the motivation of the writer. I should note that the biologists also pointed out a number of mechanical errors and stylistic problems in the writing teacher's writing.

See Charles Cooper and Lee Odell, "Considerations of Sound in the Composing Processes of Published Writers," *Research in the Teaching of English* 10 (1976): 103–15, on the influence of such considerations on professional writers.

recognize it, so Crews changes *low temperature dormancy* to *hibernation*. A reader criticizes Crews's use of the term *therapy*, which implies he is doing the lizard a favor with these injections of hormones; Crews substitutes the more neutral term *treatment*. Both authors are cautious with neologisms, so Bloch, having apparently coined the term *foward complementarity*, changes it to *reverse complementarity* when a reviewer is confused. Both authors correct, with the help of their readers, dangling participles, faulty parallelism, and the like, though neither they nor their readers would identify these errors by these names.

The important studies of funding decisions by Cole, Cole, and Rubin (see note 1, earlier) take the applicant's relation to the discipline, the status in the research community, as given, as already determined by institution, publications, citations, and previous funding. And the writer cannot do much, in writing a proposal, to change these facts, the most powerful arguments for his or her competence. But the tone of almost every sentence of a proposal can be revised to show that one is cautiously but competently scientific. Often, because of the contradictions of self-assertion in scientific prose, the most effective means of defining one's place is understatement, toning down, not one's claims for one's research, but one's language. In an earlier draft Crews questioned the received idea that "courtship behavior . . . is dependent on androgens"; later he rephrases this idea as, "courtship behavior . . . might depend on androgens." He must be particularly careful about claims of priority. He changes "the implications of this observation have been unappreciated" (which suggests that he was the first to grasp these implications) to " . . . have not been fully appreciated" (which only suggests that there is more to say about them). Asked about this change, he says that the assertion of "total originality" is "sure death" with the review committee. One of the ways he defines his place in the community is by his choice of research animal, so he must be extremely cautious on anything relating to this choice, even in apparently innocent comments on lizards. He changes the phrase "More is known about the green anole lizard than about any other reptile," which could only tempt fans of other species to object, to "A great deal is known. . . ." He must be especially cautious in using the findings of other fields outside his area of research, for instance, those of clinical research on humans. He adds the cautious note to the statement that "sexual experience appears to be the most important factor" in human sexual function, because he thinks a more definite statement, though supported by his reading, "could have gotten nailed."

Bloch also strengthens his argument by backing off from his claims, in ways that are more interesting rhetorically than scientifically. One ratio is followed, in the first version, by "We proposed that. . . ." The ratio was questioned by some reviewers of the article; the explanation of it in a later version begins, "One interpretation would be that. . . ." One of his bolder objectives in the first proposal was to "determine, if feasible, the rates of evolutionary divergence and . . . approximate time of synthesis." But this was criticized by a panel member as a "notoriously difficult" project. The later version says, more cautiously, that he would "use the reconstruction as a guide in studying the early evolution of the coding mechanism," and he refers to "the distant goal of reconstruction." In general, later versions present the interpretation suggested by his model as one hypothesis among several others.

The revisions do not, however, show that the meek shall inherit the grants. As both authors temper their claims, they also assert their authority in their specific areas of research and point to their previous accomplishments. Often this change means just a shift from passive to active voice. Crews changes "mechanisms are revealed" to "I have been able to reveal," and "New light will be shed. . . ." to "I will shed new light." Similarly, Bloch adds paragraphs on data gathered "using a program written in this laboratory." He changes "the finding of increased numbers of homologies" to "our finding, in nearly half the searches. . . ." This change emphasizes the success of the project so far and emphasizes what his own lab has contributed, even though it has not been funded to do its own sequencing (the experimental determination of the order of bases on the nucleic acid). As part of this self-assertion the writers sometimes go out on a limb. Crews adds the loaded phrase "I predict that . . ." before a claim, showing that his hypothesis is, in Karl Popper's term, falsifiable. Apparently this risky language is expected at certain points; Bloch's proposals are full of such explicit predictions and are praised for being "testable."

Perhaps the most powerful component of self-presentation is the tone of the proposal, the persona the author creates in stylistic choices. Tone is not easily traced in textual terms, but clearly both authors are concerned with sounding scientific as well as being scientific. For example, Crews explains a change from "highlighted" to "shed new light on," which was mystifying to me, by saying that the first expression was "too catchy—sounds unscientific." Bloch makes a change in tone when he refers to the object of his search as "an early precursor to both molecules," tRNA and rRNA, rather than as a "primordial molecule," a formulation he had used earlier which suggests more strongly his concern with the origin of things. Interestingly, they both allow themselves to vary their subdued tone when revising sections on the implications of their research. Bloch ended the last version of this proposal with a paragraph on broadly suggested "spinoffs." Crews added to his introduction a paragraph of data on the effects of the stress of concentration camp life on women's menstruation cycles, data he had used in an earlier letter showing the relevance to humans of work on environment and sex hormones in animals. As he explained it, this addition, with its social and emotional weight, was made to support his technical argument. "I wasn't going to use it," he said, "because everybody uses it, but when I reread it, I saw that it was making a valid point about *extreme* stress."

The first major section of the application for an NIH grant must show the significance of the research proposed. But *significance* only has meaning in relation to the existing body of literature of the field. Thus there is a tension in defining one's claim; it must be original to be funded, but must follow earlier work to be science. These writers find their place in the community by making their texts fit in in two ways, with their citations and with their terminology.

In both writers we see a rhetoric of citations, though they use these citations in different ways, Bloch usually demonstrating his familiarity with the latest work in the field, Crews highlighting his own contributions. Bloch does cite his own articles, at whatever stage of review they have reached as he writes, and he attaches a manuscript as an appendix. As he accumulates data, he is able to refer more often to his own studies. He does not usually cite authors to refute them, but to show that he is aware of parallels and contributors of data to his own work. Neither does he cite articles to establish a theoretical base, an authority for his own approach; the only major cited contribution to his method is a program and a database from Los Alamos. Many citations are tactical. The most hostile referee of an early version of one of Bloch's articles (examined in detail in chapter 3) compared Bloch's model to that of Manfred Eigen, and the editor of the journal that accepted it compared the model to that of W. M. Fitch. Bloch cites Eigen and Fitch, both major figures in the study of molecular evolution, in a later version of the proposal, taking the opportunity to show the difference between their approaches and his. This strategy seems to have paid off; one panel's summary of an intermediate version of the proposal says, "The authors have considered alternative explanations and designed their analyses accordingly."

When Crews adds citations to those in his early draft, they are usually to his own work. For instance, he expands his assertion that estrogen "plays a critical role in yolk deposition" into a two-stage description of the depletion and production of vitellogenin, bringing in more references to a successful line of previous work. This sort of change is not made just to display his productivity; his output is obvious enough from the five-page list, required by the proposal format, of publications by his group that are related to the grant. He is known mainly for his laboratory and field work, and he cites this work to support certain theoretical views, he says, "to make a point, to associate myself with these perspectives." There are risks to this approach; a critical referee of one of his articles notes disapprovingly that most of the data supporting the theory are his own. But this may just show that the rhetoric of citations in a review article, which claims to speak for the entire research program or subspecialty, must be more circumspect than that of a proposal, which is expected to give some coherence to one's own previous work. Crews's problem as an established researcher is, then, the opposite of Bloch's as a new researcher; he must interpret his empirical work to associate himself with a new theoretical line, whereas Bloch must present his untried theoretical approach as potentially productive of new data. One cites himself, one cites others, but both are trying to insert new work into an existing literature.

The addition or deletion of terms with meanings or connotations specific to a discipline may be another, more subtle way of indicating one's place in the community. I have noted that both scientists cut jargon wherever they recognize it, but they also add or change some loaded terms. A reviewer of Bloch's earliest proposal says, "Most laboratories that *do research* with either tRNA or rRNA are already analyzing not only homologies but *real* structure-function correlates" (emphasis in the original). The implication is that Bloch's researches are not research (perhaps because he is using published data), that his correlates for molecular structures are not real (because they are selected to study evolution, not to study the biological functions), and that, as the reviewer continues, "the homology results are an offshoot of the main business." After this, if not because of this, Bloch is careful to relate his homology work to the "main business" of sequencing research, to account for the possibility of convergent evolution (which would fit better with this "main business"), and to use prominently the word *function*, even though origin, not function, is his main concern.

In the latest version of the proposal, Bloch makes another significant change in terms; for almost every occurrence of the word *homologies*, the central term of his project, he substitutes *matching sequences* or some equivalent. As we shall see in studying his articles, he stumbled onto the problem, common in interdisciplinary work, of a term that has a more restricted meaning in one field than in another. In molecular biology, as the reviewer's usage quoted in the previous paragraph shows, the word indicates any structural similarity. In evolutionary biology, the word can indicate only those structural similarities that result from common origins. If Bloch used the word in this sense while trying to *prove* common origins, he would be begging the question. But precision is not all that is at issue here; the change is part of the consensus-making process of proposal-writing. To use the word in the more restricted sense it has in evolutionary biology rather than in the broader sense it has in molecular biology is to acknowledge, or assert, that one's work will fit into both disciplines.

Crews's many changes in diction suggest how meanings may vary between members and nonmembers of a discipline. Thus minor revisions improving precision can be seen as part of the adaptation of the writer's style to the literature of the discipline. One zoologist finds his use of *cycles* rather than *phases* jarring in a certain context, and she draws a distinction between them; Crews responds by changing his terminology throughout. She also points out the vagueness of the phrases *behaviorally inactive* and *sexual behavior* to an ethologist who must observe and categorize these activities. Crews substitutes phrases that have more specific meanings to an ethologist: *non-courting* and *courtship and copulatory behavior*. One of his changes shows, like Bloch's deletion of *homology*, the lines between disciplines or approaches. I had interpreted his substitution of reproductive *processes* for reproductive *behavior* as an attempt to describe his comprehensive approach more accurately by using a more general term. In fact, he says, the change is tactical. He believes that studies of behavior, especially field studies, are not being funded by NIH, whereas studies of physiology (which are what the words reproductive *processes* imply in this instance) are more attractive to them. What seems a minor revision relates to the changing fortunes of that notoriously loaded word *behavior* through the 1970s, and indicates the researcher's keen sense of the connotation of the word in various disciplines.

Changing Strategies of Presentation

We have seen that many of these writers' revisions affect their personae as researchers and relate their work to the literature and the discipline. If we look at successive versions of one short but crucial part of the proposal—Bloch's "Abstract" and Crews's "Specific Aims"—we can see how in the processes of writing and rewriting the writers respond to and develop consensus in the field. (See the texts in Appendix 1.) These carefully composed sections are the writers' chances to present the main purposes of their research programs without burying them in detailed methods and data; some reviewers may not read the rest as carefully, especially if the proposal is as long as Crews's. We shall see that, late in the revision process, Bloch and Crews come to opposite strategies of self-presentation. Bloch tries to play down the more "speculative" theoretical aspects of his program and emphasize the data he has collected so far, whereas Crews decides at last to emphasize the larger and more controversial implications of his study. Each shows, in his last version, a closer fit between his work, as he presents it, and his discipline, as he presents it. Both strategies shown in these processes of revision are attempts to deal with increased competition for health-related research funds by relating the proposed work to the consensus in the field.

We have seen that Bloch was criticized by reviewers for being too committed to his model, for being too speculative, and for wandering from the "main business" of structure–function studies. In the three versions of his abstract, we note that the model is first played down and then finally removed, that the accumulation of data is emphasized more than the larger implications, and that alternative explanations for the matches, including function, are given more consideration. The revision of the opening sentence reflects this change in strategy. In the first version he says, "A search is being conducted for sequence homologies"; the subject of the sentence is the author's action, and the tone, as in the last sentence ("An attempt is being made . . .") sounds merely hopeful. The opening of the second version is at once more impersonal and more confident; there he presents data from the research so far as posing a striking problem requiring solution: "Ribosomal RNA is peppered with tracts that are homologous with regions found among different transfer RNAs." The lively verb "peppered" suggests that these data are too insistent to be overlooked, and the reference to "different transfer RNAs" suggests a broad scope of data. In the third version this lively but still vague statement is replaced by a statement suggesting comprehensive and quantifiable findings from many species: "A large minority of tRNAs from all species of organisms studied have stretches whose base sequences are identical or nearly so to stretches found in rRNAs."

Bloch's accommodation of the discipline and his presentation of his work in terms of its consensus are apparent also in his revisions of organization and sentence structure. In the first abstract, the model occupies the central and longest paragraph. He immediately states that he is looking for evidence of common origins, not just explaining homologies; here he tips his hand and lets his critics see his larger program. The rest of the abstract is organized, logically enough, by the researcher's effort: theory, model, predictions. One methodological problem is evident in the gap between sentences 7 and 8; his data are on existing rRNA and tRNA, but he applies them to what he calls "primordial RNA." The fact that the data and the hypothesis must remain in separate sentences suggests he has not yet found the syntax to make the connection. This gap will prove to be important to reviewers.

The structure of the second version allows Bloch a longer discussion of the homologies (sentences 4–7) before he presents the model used to explain them; the focus is on the matches, rather than on the researcher and the theory, until "our work" in sentence 6. Now he mentions function as a possible alternative explanation for the homologies, and offers a test for convergence to determine its role. Still, he can only say at this point that this complicating factor "cannot yet be ruled out." The description of the model and prediction is tightened (sentence 10), giving it fewer words and less emphasis. The gap in the first version between present-day data and primordial hypothesis is not bridged but eliminated; here it is clear that the model only predicts homologies in present-day tRNA and rRNA, and needs no inferential leap into the past.

In the third version of the abstract, the model is not mentioned explicitly at all, though it is still implied in his analysis of the homologies. This version is organized by the sequence of ideas, rather than by the narrative of the researcher's efforts; it offers a sort of theoretical flowchart. Bloch says, more cautiously than before, that the homologies might be due either to function or to common origins. If function is the explanation, it might be either on the DNA or on the RNA level, and if origin is the explanation, it might be the result of either primordial or relatively recent conditions. Now the assertion of the ancient origin of these homologies is in the passive, and is after the data, so that the data, not Bloch, suggest it. The potentially troublesome statement that he is searching for ancestral RNA starts with a long noun phrase that may defuse some resistance, and uses hedging verbs: the overlapping and overlays "suggest" that further identification "should permit" reconstruction. Finally, whereas the first two versions end with this prediction, the third version ends by emphasizing "the correct functions of the transcription–translation mechanism." So for the Public Health Service he emphasizes the possible health applications, which were not mentioned in the earlier NSF versions.

Crews had also been criticized for favoring a "speculative" model that is inconsistent with much of current research, but in the revisions of his "Specific Aims" section, we see a strategy different from that of Bloch, a movement toward emphasizing his controversial model. This change was made in a very late draft, after many other changes, most of them to improve readability, had led to a draft he optimistically labelled the "final final final draft." In this draft he still cautiously plays down the model that proposes dissociated as well as associated reproductive tactics. A two-sentence introduction to his general field and specific interest is followed by a two-paragraph comparison of the green anole lizard to the red-sided garter snake. The model is subordinated to the unexceptionable comparison of two species that happen to exhibit these tactics. His own methods of investigation are not stressed. The third paragraph says that the difference in reproductive tactics has implications, but leaves those implications for the next section, where they are less prominent.

In the later version Crews highlights his more controversial approaches. The safe statement, "I will continue my study of two reptile species," is replaced with a sentence beginning, "The general objectives of my research are . . ." that introduces immediately the ecological views disputed by some reviewers. Further sentences in the first paragraph emphasize his distinctiveness as a researcher, as shown by his comparative approach and his combination of laboratory and field experiments. The second paragraph, which had been organized around the comparison of two species, is now organized around two reproductive tactics, further emphasizing his theoretical framework. He highlights the definitions of the terms he has coined by putting them in separate sentences (returning to the phrasing of a much earlier draft, written before he had started downplaying the newness of his work). No specific species are mentioned yet; the lizard and the snake are introduced only in the last paragraph, as "one representative species of each reproductive tactic." His "goal is to compare the two tactics," to look for broad knowledge of mechanisms rather than just specific knowledge on one or two species. He emphasizes the "broad approach" and the search for important generalizations. The concluding sentence of the earlier version had put direct, immediately applicable findings first, with fundamental concepts in the second part of the sentence; here it is the direct findings that come after the "also," in the position of secondary importance to the fundamental concepts.

Though the two researchers follow different tactics, they both try to relate the proposed work to the consensus in the field. Bloch saw that his proposals and articles were getting more favorable review as he gathered more data and discussed alternative explanations. Thus he presents himself as a new but well-informed and cautious member of the existing RNA sequencing program, and plays down wherever he can what he feels are the controversial aspects of his project. He need not insist on the newness of his thesis; its boldness will be apparent, to anyone likely to accept it, from the striking tendencies in the data collected so far. But he is not just persuading the panel and the discipline with these tactical changes; his reviewers are persuading him in some ways as well. Since he must discuss the alternatives to his model, he becomes more involved with structure-function relations, if only to dismiss their influence here, so the context of his research is changed by the process of applying for funding.

Crews's last-minute revisions may seem to indicate a strategy of defiance of the consensus of his subspecialty, just as the previous version seems to indicate a tactical appeal through the less controversial elements of his research. But these changes may also be seen as part of a consensus-making process, one that goes beyond the boundaries of the subspecialties of herpetology and classical neuroendocrinology to include an audience of comparative biologists and evolutionary theorists. To put this strategy in more practical terms, he may have reasoned that if only about 5 percent of the proposals to this panel were to be funded, no amount of interesting new data on anole lizards and red-sided garter snakes would be considered worth funding if it just supported existing models based on other species. If he stuck to the consensus, he might not be criticized, and might even get favorable comments from the reviewers, but he wouldn't generate enough enthusiasm to get him across what the reviewers call "the payline," the priority score cutoff for funding. He would have to present a bold idea, and present himself as a researcher capable of a uniquely broad and ambitious project. He knew, after a few hostile reviews of his related article, that in taking this approach he risked a rejection if the panel was persuaded by one of his critics. But that risk was apparently preferable to cautious dullness.

My study ends with these submitted versions of the proposals, since I am interested in how the researchers write the proposal, not in how the decision to fund is actually made. But the decisions, in this case, support the researchers' senses of appropriate strategy. The "pink sheet" summarizing the decision on Bloch's application shows the study section members were intrigued by the homologies he pointed out, but were still suspicious of his advocacy of a model attributing these homologies to a common ancestor. The summary says he needs to consider critically the other possibilities, especially convergence due to function. So he has not convinced them that his work gives sufficient attention to the work being done on structure–function relations. The major criticism of the proposal, though, is that it lacks a sufficiently detailed theoretical framework, specifically an explanation for how he will relate the present-day homologies to the ancestor molecule–how he will cross that gap noted in the structure of the first summary. This too can be interpreted as an indication that Bloch still stands outside the consensus of the subspecialty; he is being told he has not demonstrated a theory that both takes into account current concepts and also allows him to go beyond the current line of work. Though Bloch's proposal was not successful, the strategy of downplaying the model and emphasizing his awareness of structure-function studies seems to have been the only strategy that would have had a chance. Bloch's comment was that he would have to "talk to them through more publications"; that is, he would have to establish himself as a known contributor to the field before applying again. And his eventual success in getting published (described in the next chapter) seems to have helped; after an article appeared in *PNAS,* he received some funding, not from a government agency, but from two private Texas-based foundations.

Before the decision on Crews's proposal was reached, the study section scheduled a site visit at his lab to observe its work. Such a visit illustrates the consensus-forming function of the proposal process. Site visits can be scheduled by the Executive Secretary of the section (the NIH administrator) to resolve differences or doubts on the panel; they are usually made in cases of applications that are close to being funded. In some cases, the fact of the site visit would indicate a seriously split panel trying to reach some sort of agreement. But Crews's interpretation is that the administrator thought that some members of the panel were just unenthusiastic about the proposal, so that it might not get the very good priority score necessary for funding by NIH under current budget conditions. If this were the case, his strategy of emphasizing the broad implications of his work was probably wise, because the panel's conception of him as an ambitious researcher turned out to be more important than their awareness of his controversial relation to his research community.

The site visit was, in a way, a second proposal, this time presented orally, with the lab itself as the most persuasive illustration. Crews prepared by going over his proposal carefully with a number of colleagues, and he planned to temper somewhat the tone of the submitted version of the proposal. He said that he didn't want them to think he was claiming to have the last word on the relations between hormones and sexuality. As it turned out, he spent most of his presentation demonstrating that his lab was capable of such a large project. He showed a very detailed notebook of experimental prospectuses drawn up by his assistants to demonstrate his careful quality control. He emphasized the lab's publications over the last five years to demonstrate it had the capacity to handle so many projects. There was no arguing over controversial theories. Though he knew beforehand that one of the visitors would be a critic of his approach, he wasn't even sure afterwards which one this was. In the visit, as in the written proposal, persona and relation to the literature and the discipline are crucial. Crews consulted with colleagues, adjusted his tone, prepared still more textual evidence to present himself as a competent researcher and as an accepted contributor to the literature, all to enable this group to come to an agreement within itself. If his proposal had been rejected because of opposition by one powerful reviewer, this view of proposal writing as a consensus-making process would be meaningless. Instead we see still more mechanisms to allow the researcher to shape his or her persona and to make the decision representative of the subspecialty as a group.

In the end, the whole evaluation process was caught up in much larger fiscal decisions. According to Crews, a modification of accounting procedures, part of David Stockman's new budget in 1985, meant that the whole cost of a three-year project like the one proposed by Crews would have to be assigned to the first year of the project, instead of being spread over the accounts for three years as in the past. This effectively cut by two-thirds the already small number of proposals that could be funded that year, and eliminated most long-term proposals like Crews's in that round. But his lab continued to operate with its NSF funding and his own funding from a Career Development award. The grant was finally renewed, for three years, and it continues to fund part of Crews's research. He applied for a five-year competitive renewal in October 1987.

In my textual analysis, I have been showing the politics involved in the smallest details of the writing of funding proposals, but the result, based as it was on funding constraints rather than on the decision of the panel, shows how the effects of big political changes reverberate through science. There may be a lesson in this for those of us analyzing these detailed case studies, doing what is called, in social science jargon, "microsociology," reminding us that the larger institutional and policy processes analyzed by earlier students of science and society continue to operate. We have conducted a polemic to show the social in the scientific, but the social and political aspects of science are all too apparent at a time of funding cuts. There is much to be learned in studies of laboratories, but at some point, as Bruno Latour reminds us, we must break out of methodologies that assume that "science stops or begins at the laboratory walls" ("Give Me A Laboratory," p. 168).

The Uses and Limits of Rhetoric in Proposals

I have argued that the proposal-writing process shapes both the writers and, to a lesser degree, the discipline. The writers, who were doing work they saw as being on the boundary of two fields, moved toward a presentation of themselves as good members of those fields, and presented their work in terms of its interest to other researchers who might tend to reject it. There is a tension in both lines of argument. As we have seen, self-presentation requires a difficult balance—not too meek, not too assertive—that cannot always be reached by studying some generalized portrait of the good scientist. The image seems to depend partly on the type of research proposed. Both these researchers decided to present themselves in ways we might not expect. The researcher who wants to verify his model of the origin of life presents himself as the skeptical servant of the mountains of data printed out by his computer program. The researcher who wants to spend five more years in painstaking studies of thousands of snakes and lizards presents himself as a theoretician studying a new conceptual framework. There is a similar tension in their attempts to present their work as interesting, for they must show that it is original and yet entirely in accordance with the existing discipline. So they use citations, or significant vocabulary, or on occasion directly claim they can make a contribution. But here too they are limited; for instance, words like *new, fundamental,* and *important* are all but forbidden, and even *interesting* seems to provoke some readers. Claims of originality are risky, and criticisms of opposing views can seldom be explicit. Both authors wrote letters defending their work against the criticisms of hostile reviewers; comparing these to their proposals one sees how careful they were with tone in the formal proposal. When decorum is no longer demanded by the proposal format and the evaluating audience, they are unabashedly enthusiastic about their projects. They did not lose the sense they had at the beginning of being in hot pursuit of the secrets of life, though in their proposals they conceal their excitement.

But I have argued that the contexts of their projects were changed by the process, even if their enthusiasm was not. Bloch, as I have pointed out, reoriented his studies to provide mathematical analyses of the possibility of function accounting for the homologies he observes. His research was no longer just a proof for his ideas on the origin of life; it was now also a fairly elaborate method for comparing structures of molecules. Crews made no such methodological changes, but he has to think, whenever he writes a proposal, about what his work can contribute to fairly distant lines of research on other species and about how his theoretical models relate to those used by most researchers. Finding conventional terms for unconventional research is not just an exercise in rhetoric—it changes the research.

Of course the success or failure of the proposal also changes the research. But funding does not always determine if a research program continues. While Bloch searched for funds, he continued to write articles about tRNA-rRNA homologies, but followed a line of work that required less money, analyzing the published data for significant patterns. He did not get a postdoctoral student with whom to develop the theory, but as we shall see he found a collaborator in, of all areas, statistical mechanics, who was interested in developing the mathematical description of these homologies. A number of researchers have responded this way to cutbacks, moving into less expensive lines of research, but not abandoning the research program altogether. When Crews was not funded by the NIH in this round, it meant some cutbacks in the lab, but he had other grants for other projects, so it was not a question of sending the postdoctoral and graduate students, and the snakes and lizards, home.

The proposal process also changes the field in a more fundamental way, by challenging the terms in which the subspecialty defines itself.[4] Both these researchers saw themselves as working at the edge of a specialty or on the border between two subspecialties; Bloch talked about "the establishment" in molecular biology, and Crews referred to the "prevailing paradigm." When the study section gives a proposal like one of these a priority number below the payline, they draw the line that marks the edge of their specialized field. When the study section approves such research, it redefines that line. To a large degree, both researchers accepted the assumptions and criteria by which

4. See Michael Callon, "Struggles and Negotiations to Define What Is Problematic and What Is Not: The Socio-Logic of Translation," in *The Social Process of Scientific Investigation*, eds. K. Knorr, R. Krohn, and R. Whitley (Dordrecht: Reidel, 1980), pp. 197–219.

this decision is made; they disagreed with the panel only about how these criteria applied to their own work. Since they were part of the system, we should ask, not whether the system is "fair" to individuals, but how it serves the scientific discipline.

Representative Conlan, whom I quoted earlier, is not alone in asking whether the peer review process "stifles new ideas." A favorable reviewer of one of Bloch's proposals concluded, "Provocative ideas are always in short supply, and there is truth to the criticism that the present funding system often fails to nourish them." But the funding system exists to select as well as to nourish, and here the powerful consensus that Representative Conlan calls "incestuous" may serve to stabilize the economy of the discipline. For example, to approve either Bloch's proposal or Crews's would be to define large new research programs, beyond what these individuals propose, to study the origin of primitive RNA through homologies in present-day tRNAs and rRNAs, or to look for relations between hormonal cycles, mating behavior, and ecological factors in a wide variety of species. Such redefinitions of a field require changes in careers and institutions, and are enormously costly in time and money. In some cases, such as the line of neuroendocrinology Latour and Woolgar have studied, such costs may prove to be worthwhile. In this case, some reviewers might argue that there is too much left to be done on conventional structure-function studies, or on hormonal studies based on the simpler paradigm, for attention to be diverted to other lines of work, even if these other lines of work turn out someday to be important. If husbanding of resources for a consistent line of work is a function of funding decisions, it is not surprising that the proposals focus on who the writers are, whether they can do what they say, and whether, if they do it, they will have much effect on other researchers, or on problems that are important to a wider public audience.

If the rhetoric of the proposal will vary with each discipline and with the writer's relation to the discipline, it is not given by some ideal list of persuasive or communicative techniques, or by an ideal scientific persona, or even by the characteristics of the project itself. Thus, the cautious tone adopted by Bloch, appropriate for his situation as a newcomer, would be disastrous for Crews, who is well established in his specialized field. Scientists learn the rhetoric of their discipline in their training as graduate and postdoctoral students, but they relearn it every time they get the referees' reports on an article or the pink sheets on a proposal. Bloch learns where his data get a good response; Crews finds how his assertions affect a researcher who works on mammals. Finally, the researchers themselves come to assume most of this knowledge of the discipline as something natural. But we need to make it explicit and conscious to open it to people outside the discipline. As I will argue, the public needs not only to understand the facts of science, but to understand the way those facts are made.

I have focused in this chapter on the most obviously rhetorical genre of scientific writing; by implication, I am saying that the rest of the process of producing knowledge can also be seen in terms of the forms of texts. The rest of this book is devoted to what might be considered later stages in the production of a knowledge claim, as scientists address various audience in various genres. To receive credit scientists must publish claims in journals where they will be read by other researchers who will cite them (chapter 3). To organize a consensus around the claim they must address the criticisms of other researchers within the core group, either informally and implicitly or, in the case of controversies that reach print, formally and explicitly (chapter 4). Scientists write popularizations to reach beyond the small circle of specialists working on related problems (chapter 5). And ultimately, these claims can become a part of the general culture, as accepted facts about nature, or may be rejected as the notions of a small group of specialists (chapter 6). It is at this last stage in the life of a fact that there can be controversy about the interpretation of biological research in other discourses, and about the significance of these interpretations for the life of the community.

Chapter Three
Social Construction in Two Biologists' Articles

The writer of a grant proposal knows exactly what he or she wants to accomplish, and knows in general the sort of reader he or she is trying to persuade. The rhetorical purpose and audience of a scientific journal article that might result from the funded research are not so clear to nonscientists: the rewards of having an article in *Nature* do not seem so definite and immediate as, say, a grant of $100,000 a year for three years. But money is not the only, or the most important, resource in scientific programs and careers. The "cycle of credit" in science (to use Latour and Woolgar's phrase) involves the conversion of one kind of credit, grants of money, into another kind of credit, the recognition that comes with publication and citation of one's claims to have established a new piece of knowledge, and then the conversion of that recognition into money and other resources. As we saw with both Bloch's and Crews's proposals in chapter 2, grants depend on earlier published claims and are meant to produce later published claims.

The potential audience of a scientific article is so broad that interaction of the sort we have seen between panel and proposal writer is out of the question. But those who have looked closely at scientific articles have shown that, like proposals, they are designed to persuade. The immediate audience at which this persuasion is directed is quite definite: before any article can reach the diffuse and perhaps distant audience of journal readers, it must pass by the immediate and definite audience of a few referees.[1] In most academic fields, and certainly in

1. On the origins of refereeing, see Harriet Zuckerman and Robert Merton, "Institutionalized Patterns of Evaluation in Science." Charles Bazerman expands on this historical perspective in *Shaping Written Knowledge*.

David Hull has a fascinating study of the review processes in "Thirty-one Years of *Systematic Zoology*," *Systematic Biology* 32, no. 4 (1983): 315–42. Hull draws on the sort of editorial records that have not been available to other studies I have seen. He examines

all fields of biology, every claim that counts, however renowned the originator, must appear in a journal that makes decisions on the reports of referees.[2]

The referees of scientific articles are abused nearly as much as the referees of football games. Almost every scientific researcher I have interviewed has an anecdote about a referee who reviewed an article of his or hers unfairly, or who required alterations that, in the writer's view, diminished the value of the article. But there is no equivalent, for a scientific article, of the videotape replay that shows whether the referee's call was correct. I would like to look at the processes of review and revision, not from the perspective of the individual researcher confronting the individual reviewer, but from a broader perspective in which these processes are part of the functioning of a scientific community. I will suggest that the procedures of review and revision of the text can be seen as the negotiation of the status that the

---

a case in which there were widespread perceptions of biases in favor of certain approaches during certain periods of the journal's history, but he does not find any evidence to support these perceptions. He summarizes this study in a review of the broader issues raised by publication and peer review, "Openness and Secrecy in Science: Their Origins and Limitations," *Science, Technology, and Human Values* 10, no. 2 (1985): 4–13.

Douglas P. Peters and Stephen J. Ceci, "Peer-Review Practices of Psychological Journals: The Fate of Published Articles, Submitted Again," *Behavioral and Brain Sciences* 51 (1982): 187–255, presents an experiment based on the trick of submitting as manuscripts the texts of articles the same journal had already published, changing only the name and institutional affiliation of the author (the journals did not use blind refereeing). Because this article appeared in a journal that has peer commentary, there is a fascinating range of responses to this experiment, most of them from editors of other psychological journals. Peters and Ceci did not change the sex of their fictional authors. One study that does attempt to investigate gender bias is Michele A. Paludi and Lisa A. Strayer, "What's in an Author's Name? Differential Evaluations of Performance as a Function of Author's Name," *Sex Roles* 12 (1985): 353–61. But their study used the responses of college students, not those of actual referees.

2. John Ziman's work has stressed the importance of review and public consensus in the authority of scientific knowledge; for a brief summary, with basic references, see *An Introduction to Science Studies* (Cambridge: Cambridge University Press, 1984), ch. 4. Linguistics is a notable exception to the requirements of publication: key papers are often available for years only as quasipublic circulated manuscripts; see Frederick Newmeyer, *Linguistic Theory in America* (New York: Academic Press, 1980), throughout, for discussion. As Newmeyer points out, the second most cited work in syntax is the unpublished doctoral dissertation by John Ross, "Constraints on Variables in Syntax" (MIT, 1967).

Many historical studies stress the importance of informal communications; for particularly rich accounts, see Rudwick, *The Great Devonian Controversy* (on geologists in the 1830s) and Pickering, *Constructing Quarks* (on high-energy physics in the 1960s and 1970s).

scientific community will assign to the text's knowledge claim. This negotiation may not directly address the claim itself and the evidence for it, but may instead focus on the form of the text. Thus, a close study of these texts may help us see one part of what Latour and Woolgar, in the subtitle to *Laboratory Life,* call "The Social Construction of a Scientific Fact." I present two articles as cases of such negotiation, showing the range of possible claims (that is, assertions of new knowledge for which the author is to be credited). I interpret the formal changes in the manuscripts as they affect the status of the claim and account for these changes in terms of the social context, the relations between the author, the editor, the referees, and the wider scientific community. I make several kinds of comparisons: earlier articles published by the same authors serve as a background of acceptable form, successive versions of the articles show the process of negotiation, and the differences between the views of editors and reviewers on one hand and writers on the other show the kinds of tensions on which the negotiation is based.[3]

The two articles are by the authors of the proposals in chapter 2: "tRNA-rRNA Sequence Homologies: Evidence for a Common Evolutionary Origin?" by David Bloch and his colleagues, and "Gamete Production, Sex Hormone Secretion, and Mating Behavior Uncoupled," by David Crews.[4] I have chosen these articles because they show the processes of social construction with particular clarity, each having had a rather bumpy ride before appearing in print. Each article had been through four reviews before it was finally accepted, Bloch sending his to *Nature* (twice) and *Science* before it was accepted in a revised version at the *Journal of Molecular Evolution,* Crews sending his to *Science* (twice), *Nature,* and *Proceedings of the National Academy of Sciences* (*PNAS*) before it was finally accepted at *Hormones and Behavior.* The authors rewrote the articles each time, so that the published versions are hardly recognizable as related to the first submissions. These texts are atypical; they mark departures in the careers of two well-established researchers, so they show the tensions operating in the process of publication. As we have seen in studying their proposals, the two authors are different enough from each other to be compared in social terms, one entering a new field, the other well estab-

3. For my use of these terms, see Gilbert and Mulkay, *Opening Pandora's Box.* John Law and Rob Williams, "Putting Facts Together: A Study in Scientific Persuasion," *Social Studies of Science* 12 (1982): 535–58, reach conclusions similar to mine in an analysis of discussions among the coauthors of articles and relate these conclusions to the concept of a network.

4. See the Reference List, section 2, for texts discussed in chapter 3.

lished in a network of researchers. And the long, drawn-out review, though unusual, helps us see the texts in evolutionary terms: we see in detail responses and decisions that are usually compressed and unnoticed even by the participants.

I collected the various manuscripts of these articles that were submitted for publication, and a few of the many intermediate drafts, together with comments of the authors' colleagues and of the reviewers, and the authors' responses to these comments, in correspondence with the editors and in interviews with me. I marked changes, however trivial, and made some guesses about why these changes were made. Then I interviewed the authors, asking them for their own interpretations of the changes, and submitted a draft of this paper to them for comments. Thus my conclusions draw on three kinds of interpretations of the texts and revisions: my own reading, in complete ignorance of biology, the comments of the authors' colleagues and then of anonymous reviewers, and the explanations given by the authors themselves. I was interested in the differences between these readings, not in determining which one was "correct." My assumption is that none of them is privileged, so I have relied on neither the authors' claims for the importance of the articles nor the reviewers' doubts. I have tried also to avoid privileging my own outsider's perspective, but that perspective may tend to dominate, just because it is the basis for my narrative.

Why did these articles take so long to get published? The explanations for the delays depend on whether one focuses on the individual researcher or on the structure of the research community. A rhetorician might say that the authors had to invent by trial and error the arguments by which they could persuade their audience to assent to their claims. A sociologist interested in how social factors distort what he or she considered objective scientific research might say that the authors' research programs conflicted with the individual interests of reviewers, who had their own established research programs; before they could publish they had to find journals of subspecialties in which their work was not a threat. Both these approaches are useful, but they both underestimate the social nature of the publication process. The rhetorical approach, in treating the problem as a matter of strategy, accords the writer more conscious control and detachment from the audience than I can see; we must remember that Crews and Bloch frame their ideas, however unorthodox, within disciplinary assumptions. The approach through individual interests, although it points out some kinds of social influences, overlooks the way the very form and language of the article tend to create consensus. It would be a mistake, for instance, to analyze the institutional affiliations of the authors or their age or class background as "social" elements, while ignoring such textual features as the footnotes or the tables of data or the partitioning of the article into sections, all of which are governed by textual conventions that shape the way a claim can be made.

I argue that the process of writing a scientific article is social from the beginning, because it involves compromises between opposing rhetorical demands and opposing goals. In the proposals we saw a tension between the need to show the originality of the work planned and the need to place it in the context of existing work in the field. In scientific articles there is a similar tension that makes negotiation between the writer and the potential audience essential. On the one hand the researcher tries to show that he or she deserves credit for something new, while on the other the editors and reviewers try to relate the claim to the body of knowledge produced by the community. But the claim must be both new and relevant to existing research programs to be worth publishing; the writer cannot please the audience just by being self-effacing. The result of this negotiation is that the literature of a scientific field reproduces itself even in the contributions of those who challenge some of its assumptions.

This tension is brought out in the negotiations over the published form of the text, so we can see, in arguments over the tiniest textual details, larger tensions over the claim, its appropriateness to the journal, and its form. The claims in these articles can be taken on several levels of significance. For instance, one of these claims may be allowed as a description of one species, or as an interpretation of a process applicable to all species, or as an argument for how this process evolved. The claims that are restricted to descriptions of the data are not inherently more scientific, or even more publishable; they are just one level of a hierarchy in which the place of the article is being negotiated. The same claim may be considered "speculative" or "well defined," a "highly significant" advance or a "well-known" observation, depending on the body of literature into which it is placed and the audience which is to read it. We can see in the referees' comments negotiations over how the claim is to be placed. In general, the authors start by making high-level claims for the importance of their findings, whereas the reviewers demand that they stick to the low-level claims that take their findings as part of the existing structure of knowledge.

Much of this negotiation over the status of the claim concerns the "appropriateness" of a paper to the journal to which it has been submitted. Of course the authors want to see their papers published in prestigious journals that testify to the importance of their claims. They also have practical reasons, because of the interdisciplinary nature of their work, to want to appeal to a broad range of researchers in other specialties who might, if interested, provide data to support their claims. And, like most authors, they want speedy publication, especially because the articles would then support the related proposals for funding that we saw in chapter 2. The referees, on the other hand, see their function as that of sorting papers by levels of importance, by subspecialty, and by genre. Their use of the word "appropriate" or "inappropriate" in evaluating an article might suggest that each manuscript is unambiguously marked as one sort of article or another, but the sorting is not, in fact, automatic. Here too we see a tension finally resolved in the compromises that allow an article to appear in print.

There is a similar tension over the form of the article. These authors have some difficulty fitting their new interpretations into the form of the research report or the review article, because these forms demand that the claim fit closely into the structure created by other scientific articles. The authors try to bend the constraints of form to fit their ideas—in effect, to tell their stories from the beginning—whereas the reviewers try to use the form to make the ideas fit into the literature as a whole. Again, they are arguing, not over the writers' failure to use the correct format, but over the type of the claim and the importance to be accorded to it.

The Writers

We have already seen that the ideas in their proposals could be considered controversial by other researchers. Some readers of this chapter might tend to accept these referees' reports and assume that Bloch's and Crews's difficulties in getting their articles published must be traceable to their own eccentricities or scientific skills, rather than to their claims in relation to the literatures of their fields. But as my brief biographies in chapter 1 show, both authors have had successful careers, and they are both familiar with the literature of their fields and with the processes of publication. The *Science Citation Index* shows that most of the cited publications of both authors are reports of experimental findings or field observations published in the core journals of their disciplines; it also shows an important difference in the authors' positions in the subspecialties to which these new articles matter. David Bloch receives about forty citations a year for articles as far back as 1954; his most cited paper is a 1969 *Genetics* review article that still gets ten to fifteen citations a year, an unusual number for a fourteen-year-old article in most fields. He seems to have written several cited articles a year through the 1960s (with fewer in the 1970s, as his interests changed), all in journals of cell biology or genetics, in general journals like *PNAS*, or in specialized handbooks. But the article considered here is his first publication in the field of nucleic acid research, a very competitive field in which most publications in the major journals come from large groups at well-established labs.

David Crews has been first author of five or six cited articles a year since 1974; with the work on which postdoctoral, graduate, and undergraduate students were first authors, his lab produces about fifteen articles a year. These papers fall into several categories: articles in journals of zoology and endocrinology, those in more general journals, such as *Science* and *PNAS*, chapters in books, and popularizations in *Scientific American* or *BioScience*. Currently, his most cited entries are articles in *Science* (1975) and *Hormones and Behavior* (1976), though a controversial report in *PNAS* (1980) received a number of citations and some news stories soon after its publication. This last article sparked off the controversy analyzed in chapter 4.

The articles I have chosen to discuss differ from these earlier publications in complex ways that I can summarize by saying that the writers each had a big idea. Most of their earlier articles, whether experimental or theoretical, had answered questions posed in the literature. In the articles I consider, though, Bloch is answering a question that had not been asked, and Crews is giving a new answer to a question that had already been answered. The articles are especially suited to illustrate the tensions I have outlined because they are interpretations of published data in new terms, rather than reports of experiments continuing an established research program. Thus, if the writers are to have any effect, they need to claim significantly new interpretations and reach a broad audience, for they cannot just contribute data to, or get additional data from, the researchers in their immediate subspecialty. The two authors have a similar problem, but their processes of publishing these claims are different, partly because of their differing positions in their research communities.

For Bloch, the big idea came when he started work in an area entirely different from the cell biology studies he had pursued for twenty-five years. As we have seen in chapter 1, he started a new line of thinking in the late 1970s after a back injury and a graduate seminar give him the opportunity to think about something other than the main work of his laboratory. In 1981 he wrote, but did not try to publish, a paper on "The Evolution of Evolution," and from then until

1986 he wrote a number of drafts of one article, and versions of several related studies. Bloch, then, is an unusual example of a writer completely new to a specialized field who is (as he must be for my purposes) quite familiar with the way biology journals in general review articles for publication.

In his acknowledgments, Crews traces his big idea to discussions with colleagues whom he thanks as "innocent bystanders" and to mentors who taught him "the value of a comparative approach." He has also pointed out in conversation how these teachers made him skeptical of "deterministic models of behavior"; he sees himself as carrying on, in his experimental studies, a form of the nature vs. nurture argument. As I noted in chapter 1, his training in several different disciplines might tend to make him receptive to unorthodox ideas. His popularizing articles and reviews require him to explain and rethink basic principles, as Bloch must do for his freshman classes. The many grant proposals needed to support his large lab, most of them to health agencies, require him constantly to justify his work with reptiles in terms of its significance for humans, so he must consider its ultimate relevance. Whatever the reason, he, like Bloch, proposed a claim that was at variance with the current literature in his field, that needed support from researchers in several fields, and that led him to try a different form from that of his earlier reports and reviews of research. He, like Bloch, based a proposal for funding on this claim, and his proposal and article evolved together. Meanwhile, he and his colleagues in the lab continued to publish other articles reporting new data from their experiments and field observations.

The differences I have suggested in the positions of the two writers may or may not affect the judgment of editors and referees. But I suggest that these differences are felt before the articles are submitted, in the writing and revision of the papers. Bloch, when he was writing this article, had no research network: few, if any of his graduate school friends, colleagues, and students worked on this aspect of nucleic acid research, and the leaders in the field did not know of his work. In this sense, he, as a full professor, was more isolated than a new graduate student, who would have, at least, a sponsor. He valued the help of students a great deal, gave them coauthor status, and remained in close contact with R. Guimares, one of his coauthors, who visited Texas from Brazil for five months in 1981. But he did not then have constant informal contact with coworkers who are expert in this area. So he did not hear arguments against his claim before he submitted a paper, and he had no day-to-day sources for new arguments to support his work, or new data that could be relevant. After years of collaborative work in cell biology, he was acutely aware of what he was missing; he wanted to get the project funded, not primarily for the equipment and time it would give him, but to have a postdoctoral student "to bounce some ideas off." His coauthor on a more recent manuscript that grew out of this project was a physicist specializing in statistical mechanics; his quest for collaborators finally took him outside of the discipline of biology.

Crews's lab, on the other hand, is an important node in his network of researchers. His graduate school training and two postdoctoral positions doing research related to his current work, his teaching of dozens of undergraduates and graduates who are now themselves teaching at other schools, his fellowships and visiting professorships, and his dozens of conference papers and lectures have given him daily formal and informal contacts with many researchers in both zoology and psychology. A recent article on which he is first author lists five coauthors at three other universities. Any article he submits for publication has been criticized by many readers; it is already the product of a community. And to a large degree, he has internalized this community; he could easily predict the contents of negative reviews (and in some cases could guess the identity of the anonymous reviewers). Crews's position in a network does not mean his articles or proposals are always accepted, but as we will see, it enables him to be considerably more flexible in the negotiation over the status of his claim, in the revision of his article, and in his choice of outlets and audiences.

Determining the Claim

Despite these differences in the positions of the authors in their fields, their two articles go through roughly similar stages on the way to publication (Appendix 2). I shall illustrate the negotiations over the published form of each article with six texts.

1. First each author wrote a wide-ranging draft he did not submit for publication.

2. Each then wrote a more limited and conventional manuscript for submission to major interdisciplinary journals: *Nature* (in Bloch's case) and *Science* (in Crews's case). Bloch's was rejected without review, whereas Crews's was reviewed by two referees who split in their decisions, and also rejected.

3. Each author then revised and resubmitted the manuscript to the same journal, with a cover letter asking for reconsideration; both were reviewed and again rejected.

4. Still confident that their manuscripts were important, they resubmitted to other prestigious interdisciplinary journals, Bloch revising somewhat for *Science,* Crews revising drastically for *Nature.* This time Bloch's article got an ambivalent but generally favorable review, but was still rejected, whereas Crews's article was returned without review.

5. After these rejections by *Science* and *Nature,* both authors submitted to journals with more limited audiences that seemed more likely to accept the articles. Bloch sent a revised version to a journal recommended by one of the referees at *Nature,* the *Journal of Molecular Evolution.* It was accepted on the condition that certain changes suggested by the referees and editor be made. Crews submitted the unrevised *Nature* manuscript to *PNAS,* where the referees were generally favorable but still recommended rejection.

6. Finally, both articles were published. Bloch's manuscript was accepted in its revised form in the *Journal of Molecular Evolution,* where it appeared in the December 1983 issue. Crews's unrevised manuscript was accepted at *Hormones and Behavior* on the basis of its previous reviews, and appeared in the March 1984 issue. The revisions between each of these stages are extremely complex, ranging from massive cuts and additions to the shifting of an adjective or a comma. I shall focus on changes that seem to affect the scope or the form of the article, for these are the features that seem most crucial in the negotiation of the status of the claim.

A citation such as "(Watson and Crick 1953)" refers to a single knowledge claim an article makes, in this case, for instance, the claim that the structure of the DNA molecule is a double helix with chains of phosphates on the outside and particular pairs of bases connecting them. Nigel Gilbert has shown how one published article may contain a number of possible knowledge claims, from which the authors and readers select the claim relevant to the model by which they are interpreting the article. Latour and Woolgar have arranged the interpretations of the claims in an article in a five-level scale of statements from "fact-like status" to "artefact-like status." They show how statements can be transformed from one type of statement to another by addition or deletion of "modalities," statements about the statements, as in "The structure of GH.RH was *reported to be* X." Trevor Pinch also proposes a hierarchy of claims; his is arranged in terms of what he calls increasing "externality," from the lowest level, statements about the observing apparatus ("Splodges on a graph were observed") to the highest level, statements about phenomena at several removes from the observing apparatus ("Solar neutrinos were observed").[5] He points out the similarities between these poles and the philosophers' opposition of claims with high veracity to claims with greater theoretical significance. I see a hierarchy similar to these in the two articles I am describing, but I prefer to define it in terms of the distance between the authors' claims and the claims of the particular part of scientific literature in which they are to be placed. The issue is the way the claim fits in what Pinch calls the "evidential context." The higher-level claims, in each case, involve contradiction of large bodies of the literature, of claims that underlie many research programs or claims that are particularly well entrenched. The lowest-level claims contradict nothing, but neither do they add anything to what has been accepted. Like Pinch, I see this hierarchy as determining, not just the degree of acceptance or rejection of a particular claim, but which claim is accepted or rejected. Like Latour and Woolgar, I am trying to base this hierarchy in the language of the claims rather than in some inherent reliability or unreliability of methods of observation or experiment.[6]

As Pinch points out, higher-level claims are likely to be profound but risky, whereas lower-level claims are likely to be taken as correct, but are also likely to be trivial. Both the biologists I am studying try to make the highest-level claim the editors and reviewers will allow (Appendix 2). Bloch's highest-level claim appears only in the early draft he did not submit for publication; he identifies a fundamental concept that he says links several kinds of evolution. "Transfer of control . . . given the name 'surrogation,' marks the appearance of new kinds of behavior at every level or organization and process, including evolution itself." His first manuscript submitted for publication just presents the model and makes the claim that "a primordial tRNA produces through successive rounds of elongation a molecule with multiple functions of gene, message, and scaffolding, and which serves as a source of the original tRNAs and rRNAs." Supporting this model, in the same manuscript, is a more limited claim, an interpreta-

5. Nigel Gilbert, "The Transformation of Research Findings into Scientific Knowledge," *Social Studies of Science* 6 (1976): 281–306; Latour and Woolgar, *Laboratory Life*, pp. 78–86; Trevor Pinch, "Towards an Analysis of Scientific Observation: The Externality and Evidential Significance of Observational Reports in Physics," *Social Studies of Science* 15 (1985): 3–36; Harry Collins, *Changing Order*, has a discussion of these approaches to presentation of claims in chapter 6. Bruno Latour discusses the modalities at greater length, with a variety of examples, in *Science in Action*, pp. 21–44.

6. I discuss the form of these statements in terms of Brown and Levinson's linguistic analysis of politeness in "The Pragmatics of Politeness in Scientific Articles," *Applied Linguistics*, 10 (1989): 1–35.

tion of data: "The patterns and distributions of homologies make phylogenetic relatedness a more plausible explanation than evolutionary convergence." This is the major claim that remains in his first revised version of the article after the reviewers' criticisms. A still more limited claim shows the relation on which the interpretation of a common origin for the two molecules is based: "The existence of homologous sequences among tRNAs and 16S rRNA is demonstrated." This is the claim that most interests the reviewers, who generally agree in finding his data on matching sequences intriguing.

It would be possible for Bloch to make an even more limited claim, stating the sequences of the RNAs without insisting on the homologies, but since he is using already published data rather than doing his own sequencing, such a very limited claim would not be publishable. Even the observation of homologies is trivial, in his own view, without some explanation of why this pattern should be noticed. So we cannot simply say that Bloch should avoid speculation; he has to try to make one of his higher-level claims stick. If the model for the evolution of RNA is accepted, he will have one piece of his larger argument in place. He may have selected this particular piece because he can define the claim for the homologies clearly and design a research program to support it using computers (to which he has access) but requiring no new (and expensive) equipment. This awareness of what constitutes a "well-defined" claim and a practical research design are part of what he brings with him from his cell biology work; he doesn't have the contacts, but he does know the conventions.

Crews, like Bloch, makes higher claims for the implications of his work that are supported by lower-level claims interpreting his observations and still lower claims showing what he has observed. His highest claim, the claim that relates to nature vs. nurture arguments, is that environmental factors may influence the evolution and development of three aspects of reproduction: "(i) The functional association among gamete production, sex hormone secretion, and mating behavior, (ii) The functional association between gonadal sex (= male and female individuals) and behavioral sex, (iii) the functional association among the components of sexuality."

His first submitted manuscript limits the claim somewhat by focusing on the first of these aspects, the assertion that, contrary to the assumption of what he calls the prevailing paradigm, these processes can be dissociated. Supporting this claim is the more limited claim that there exist many species in which gamete production, sex hormone secretion, and mating behavior *are* dissociated, and that these species need to be studied further. This is the claim that the more favorable reviewers emphasize as an addition to the structure of claims in the literature. When Crews revises his claim to this level, he loses some of the argument for ecological approaches he was making in his first draft; the claim for dissociation can be made at this level without any reference to the environmental factors leading to such dissociation. Supporting this claim is the still more limited claim that these processes of reproduction are certainly dissociated in at least one species, the red-sided garter snake. This last claim is accepted even by hostile reviewers, but Crews has already published his studies of garter snakes, so to make that claim alone would be trivial. It belongs, not in *Science,* but in the popular article he and William Garstka wrote for *Scientific American* (see chapter 5). Crews, like Bloch, must limit the scope of his claims. But as we shall see, he has the advantage that his claims, however contrary to established research programs, emerged from those programs and can be related back to them.

The hierarchy of claims has some relation to the hierarchy of journals to which Bloch and Crews submit their articles, at least in the case of some of the most prestigious journals, which insist that the claims in articles they publish be of interest beyond any one subspecialty. Bloch's decision to send his article first to *Nature* and Crews's decision to send his first to *Science* indicate how important they considered their claims, since the editors of these journals say in their instructions to contributors that they select "items that seem to be of general significance" (*Science*) or "reports whose conclusions are of general interest or which represent substantial advances of understanding" (*Nature*). Neither publication is limited to biology articles, but since there is, apparently, no biological equivalent of *Physical Review Letters,* they fill the role of rapid-publication, prestigious journals. And they offer the access to the broad audience both authors need if they are to find a wide range of data to support their hypotheses.

The articles were rejected, originally, on the grounds of the "appropriateness" of their articles to these journals, rather than on the grounds of faulty interpretation of data. The editor of *Nature* returned Bloch's article without review, saying that the journal was unable to publish manuscripts that, "like yours, are very long and speculative" and suggesting he send it to *The Journal of Theoretical Biology.* The words "long" and "speculative" and the alternative journal suggested place the article in the hierarchy of claims: the editor does not accord the claim in this form the status that would justify such broad implications, so much space, or such a broad audience. Bloch's response, in a covering letter with a revised manuscript, shows he reads the editor's criticisms as part of a negotiation, not just as a formal criticism of the length of the article. But his response also shows he is not yet willing to give much up by making major revisions. He argues that though the article seems speculative to the editor, it is in fact "an analysis of hard data" that "makes predictions" on the basis of the model, and "these predictions were fulfilled." He argues that the speculation at the beginning "is appropriate" and "would be conspicuous by its absence." If it seems to belong in a more specialized, more purely theoretical journal, then the editor is overlooking "a completely new slant on the origins of RNAs and of coding mechanisms . . . a 'Rosetta Stone' for the origin of life." He apologizes for seeming to make "extravagant" assertions of his claim, but such assertions are the only way he can press a claim not grounded in the literature.

This letter from Bloch asking for reconsideration suggests he saw the rejection by *Nature* as an oversight on the part of one editor, not as part of the community's assessment of his claim. But when his revised version was given to referees (three rather than two, suggesting the editor tried to resolve some ambivalence), they made comments similar to those given by the editor, focusing on the status of the claim rather than the evidence or argument for it. As a high-level claim about the origin of RNA it lacks rigor; as a low-level claim making some observations about homologies in RNAs it lacks general interest beyond the subspecialty concerned with molecular evolution. One referee suggests the claim cannot be supported by the literature of molecular biology, and thus belongs in what he sees as a less rigorous subspecialty: "The manuscript drifts into unsubstantiated speculation; this, however, is common in evolutionary papers." Another shares these doubts about the "highly speculative evolutionary model." Two of the reviewers suggest it be sent to "a more appropriate journal" or "a more specialized journal." The other suggests *Nature* could "publish a much briefer account of the homologies together with a brief possible interpretation of it." Although two of these readers raise statistical questions, and one refers to an earlier article, famous in the field, that puts forth a similar idea, none of them attack the evidence so much as they question the status of the claim itself. The reviews at *Science*, although much more favorable, also deal with the status of the claim, splitting the higher-level claims from the lower-level claims: "It is not clear that the empirical observations of homologies and the discussion of pmfRNA [that is, the model for common origins] can both be adequately presented in a single paper which meets *Science* page limitations." The editor apparently took this tension between claims as unresolvable, for she rejected the article without suggesting any rewriting.

The comments on the claim by referees at the more specialized journal that accepts Bloch's paper, the *Journal of Molecular Evolution,* are actually quite similar to those at *Nature* and *Science,* but here these comments do not indicate that the paper is inappropriate to the journal. "The hypothesis of this paper is of interest to evolutionists," one begins, suggesting that Bloch has found his niche in the hierarchy. Although by the *Nature* referees' standards the hypothesis is quite far from the data, here the referee finds claims that are, in Popperian terms, falsifiable: "Both the hypothesis and the data are clear-cut enough so that if the authors are wrong they will hear about it quickly from other scientists." But even the most enthusiastic reviewer is concerned that the article, while significant, does not entirely fit in the current structure of claims. If the statistics check out, the reviewer says, "This is an important finding that needs to be explained," and he or she "strongly recommends" publication. But before Bloch can make higher-level claims, his lower-level claims must be accepted by the rest of the research community of the subspecialty. "The essence of an initial paper should be to document the reality of the homologies rather than extensive studies of their origin. If such discussion is to be included at all, it should be far more balanced and less speculative." In these terms, his discussion is "speculative," not so much because it runs ahead of the data, but because it runs ahead of the literature.

The criticism of Crews's first manuscript by the referees at *Science* also focuses on the placement of his claim. One sees it as placed too low on the hierarchy, in relation to the accepted knowledge of the field: "I learned very little . . . the model is really very simple." But he can imagine a more important article on the same topic and by the same author that would be appropriate to this journal. "I am ambivalent. *Science* needs some articles in important areas such as this. I think the author, who has done some very important work, can do a better job of putting things together. . . . *Science* is read by such a wide audience that this article will certainly reach the audience that needs it most." This ambivalence makes sense if we see that the decision being made concerns not just the acceptance or rejection of the article, but also how the knowledge claim will finally be presented to the community. The status of his claim is indicated by a number of formal features that can be negotiated separately; the referee can choose to accept the author's claim without accepting some aspects of its form. The other, entirely negative review, also places the status of Crews's claims. The referee locates three claims Crews is making: one is "an accepted fact," (that is, a claim at the lowest, trivial level), another is "not a new or startling observation" (also at a low level), and the third is "a quantum leap from faulty premises" (a higher level than the referee will grant). He sees, as does the more positive reviewer, that the negotiations here concern the status that this journal can confer upon the claim: publication of this material in a journal as prestigious as *Science* "could set the field back by providing a strawman for those that feel it necessary to refute the thesis." In effect, he doesn't say, "it's wrong," but says, "it doesn't belong in this field."

Crews's letter to the editor in response to the *Science* criticisms shows that he, like Bloch, is aware that formal points are part of a negotiation of the status of his claim. Like Bloch, he defends his claim in a language quite different from that of the article itself. But unlike Bloch, who can only point out to the editor the importance of his claim, Crews is enough a part of the network of researchers to be able to fit his claim into its structure of knowledge, or at least to try. His originality, he says in his letter, is not in the observation but in recognizing its larger implications. Even if "this observation has been around for at least forty years, its significance at the conceptual level has been unappreciated." Whereas the reviewer says his claim is "an accepted fact," he can show that the opposite view is held by the standard textbook (from which he quotes) and in two recent *Science* articles. He implies that his refutation of this view is appropriate to the same journal and audience.

But in a second review at *Science*, the referees are even further apart than before. One referee says again that the claims are either well known already ("Only the naive who had done no reading would suggest that . . . no one has ever claimed that") and out of touch with current knowledge ("it is entirely different from the well-documented findings"). Finally he or she questions whether the paper says anything definite: "I was unable to find this experimentally testable hypothesis, as were two of my colleagues who I asked to read the paper." The second referee seems to have read a different article from the first: "The author of this paper provides a valuable service . . . a needed jolt . . . another important contribution . . . a clever and reasonable hypothesis." Although the referee says "any biologist with even a passing interest" in the topic is aware of some of the specific instances Crews cites, he or she sees the usefulness both of a list showing "a large number of such exceptions" and of his evolutionary hypothesis, which "will certainly generate debate and further research." In a sense, the two referees did read different articles. Crews suggests that the first reviewer is a classical neuroendocrinologist, and the second a comparative zoologist. If that is the case, they are placing the claims of the article in different hierarchies, so different that where one reader finds a clever and reasonable hypothesis, the other finds no hypothesis at all.

The comments on the shortened manuscript Crews submitted to *PNAS* show that it is not necessarily enough that the claim of the article be significant; it must have the right sort of significance. Although the referees grant the importance of the article, they both say that it does not make the kind of claims appropriate to this journal. *PNAS* is certainly a prestigious journal, but the checklist it sends to referees to get their comments suggests that, unlike *Nature* and *Science,* it selects claims presenting new data that fit an already established conceptual framework. So one referee says that "this article breaks new ground" but decides that "While I found the hypothesis as presented quite interesting and worthy of serious experimental attention, I do not think this idea merits a separate *PNAS* article." The other review shows clearly how the status of the claim may be separated from the question of appropriateness to a specific audience and journal, so I shall quote it at length.

> I have little problem with my recommendation regarding this paper: it does not belong in *PNAS.* I think that the points raised are of great importance, that the scholarship is genuinely profound, that the conceptualization is original, that the presentation is crystal clear and not obfuscated by unnecessary information and argumentation. I would strongly urge its publication in a more general journal (obviously, I would think first of *Science* or *Nature*) where it will receive the attention it deserves. . . . It is because I consider this survey/thesis to be highly significant that I do not think it belongs in a journal that publishes "data" papers.

Though it may have struck Crews as ironic that his paper would be rejected for being "highly significant," and that he would be referred back to the journals he had spent months trying to satisfy, this reviewer's decision makes sense. It is consistent with earlier reports in focusing on the "issue of appropriateness," and on determining just what kind of claim is being made, rather than evaluating the evidence for the claim. So for this reviewer a "highly significant" theoretical formulation is as much out of place in *PNAS* as would be an article with weak data or unimportant claims. This interpretation of the reports is confirmed by the decision at *Hormones and Behavior* (which often publishes the work of Crews's group—three articles in that issue alone), for Crews just sent that journal the reports he had gotten at other journals, and the article was accepted without further review or revision. The editors seem to have accepted the favorable reviews I have quoted and to have discounted the unfavorable comments, which dealt with the article's relation to other specialties or its appropriateness for more general journals. So the same placement of claims that was grounds for rejection at *Science* and *PNAS* is taken, at a journal devoted just to one specialty, as grounds for acceptance.

What has changed in the course of these negotiations? Both Bloch and Crews have altered their claims, choosing a somewhat more limited claim before submitting the manuscripts for publication, and then cutting their more controversial, higher-level claims in their revisions. In exchange for publication, they accept a different level in the hierarchy of claims. They have also settled for less prestigious journals and more specialized audiences, accepting for these claims a somewhat different status than that they had first proposed.

Choices in Form and Style

So far I have described only those referees' comments concerning the statement of the claim itself and its appropriateness to a specific journal. But referees' comments about such matters as length, organization, and style are not just matters of taste; they too help define the status of the claim. As there is a tension in determining the appropriateness of the claim for a particular journal, between assertions of originality and participation in an established structure of knowledge, there is a tension in determining the form of the article, between the construction of the idea as the author tells it and the conventional formats of the report or review article, which emphasize the placement of the article within a body of literature. As Bazerman and others have shown, these formats, though flexible within limits, embody the attitudes of a subspecialty toward claims, methods, and use of the existing literature. And the conventional tone of scientific articles carries assumptions about the appropriate persona for the researcher. The author has a story which he cannot tell as his own, ignoring the literature, and yet does not want to fit completely into the format, distorting the shape of his idea.

A number of writers (including most notably Peter Medawar) have commented on the differences between narratives of the actual experience of science, with all their odd sources of ideas, wrong turnings, and unexpected discoveries, and the presentation of science in journal articles, the form of which suggests a method of pure inductive logic. Thus it may seem strange for me to speak of the author's "story" in describing the forms of these articles, as if they had presented their ideas in autobiographical fashion. But Latour and Bastide have shown how some narratives can remain within the form of a research report, latent in the methods section or in descriptions of physiological processes. And Mayr and others have commented on the particular importance of narratives to biological argument, which, unlike the physical sciences, must often deal with unique events in time: "Explanations in biology are not provided by theories but by 'historical narratives.' "[7] These narratives need only be implied in most articles, so that, for instance, observations of successive generations of Crew's *Cnemidophorus* can make sense without a retelling of the narrative of genetics, and Bloch's observations on sequence homologies do not need to be supported by a history of the breaking of the genetic code. Perhaps Bloch and Crews use somewhat unconventional forms for these articles because they find they need to retell a whole narrative from the beginning, rather than dealing with just one incident within the narrative given by the scientific literature. In these terms, each deviation from what the editors expect may be, not an error, but an assertion of the status of the claim, of its originality. The choice of form suggests the audience that the author thinks the article deserves. In the simplest example, an editor will not allow an unusually long article unless he or she considers it unusually significant. The reviewers' comments suggest that a similar kind of evaluation is made whenever the organization or tone of an article departs from the conventions. As Bloch and Crews gradually move from the somwhat unconventional forms of their earlier manuscripts to the more conventional versions that are finally published, they are accepting the status these referees accord their claims, accepting the decision that their claims do not call for special formal treatment.

The form of the earliest draft by each author reflects the route he took to this research program. Though Bloch's early draft, "The Evolution of Control Systems: The Evolution of Evolution," apparently follows the format of a review article, with an abstract, introduction, definitions, examples, and copious citations, the style is personal and exploratory, allowing for digressions (labeled as such), asides, suggestions of possible lines of thought left unexplored, and references to a wide range of authors outside the subspecialty, from Darwin to

7. Bruno Latour and Françoise Bastide, "Writing Science—Fact and Fiction: The Analysis of the Process of Reality Construction Through the Application of Socio-Semiotic Methods to Scientific Texts," in *Mapping the Dynamics of Science and Technology*, ed. M. Callon, J. Law, and A. Rip (London: Macmillan, 1985), pp. 51–67; Ernst Mayr, *The Growth of Biological Thought: Diversity, Evolution, and Inheritance* (Cambridge, Mass.: Harvard University Press, 1982), pp. 71–73.

Delbruck to Prigogine. The evolution of RNA, the topic of the article he will submit for publication, is here just a one-page example of the genetic code. Only at this early stage do we see in the text the relation of this model to the ideas of code, information, control, and culture with which Bloch began his thinking. Only a reader of this draft, or someone who had had a chance to hear Bloch talking informally or at a poster session, would suspect that his real goal was an explanation of the origin of life.

Crews's early draft, titled "New Concepts in Behavioral Endocrinology," also shows more of the relation of this research to his larger thinking than does the first version submitted for publication. The paper seems to have been written for people who are already receptive to his ideas, terminology, and criticisms of current concepts; the eight names in the acknowledgments suggest that only a close group of colleagues had read it yet. For this audience, he can safely follow an organization that is more exploratory than argumentative, opening with broad questions, making a leisurely review of his recent work, and only presenting his alternatives in the last pages. At this stage, one can still see the relation of his research on dissociation of gamete production and hormones to his larger assertion of the importance of environmental factors in all aspects of the evolution and development of reproduction. Crews's first draft, like Bloch's, is closer to the form he uses for oral presentations than to that of his other articles; it lacks only the slides with cartoons of lizards.

The manuscript Bloch sent to *Nature* is much more conventional than his early draft. But we can see in it a tension between the form of the report on research and the more exploratory form he had given up by comparing the submitted draft to a recent article that be had written with colleagues on cell biology, "DNA and Histone Synthesis Rate Change During the S. Period in Erlich Ascites Tumor Cells." In the *Nature* submission, six pages of introduction provide the reasoning behind the model, and then just two pages of methods and results describe the work, before nine pages of discussion and two pages on "Further Evolution." So the article is about 32 percent introduction, 10 percent methods and results, and 58 percent discussion. In contrast, the cell biology article is about 11 percent introduction, 47 percent results and methods, and 42 percent discussion. If Bloch's problem is that he is answering a question that has not yet been asked, his solution here is to start in his introduction with the most fundamental questions—the conditions for the first protein synthesis—and work toward his interpretation. In cutting the methods and results section so drastically, he may be assuming that his extensive tables (with forty of his sixty references) are striking enough in themselves to attract attention, and too straightforward to need explanation. The bulk of the paper is in a discussion titled, "Common Descent, Evolutionary Convergence, or Coincidence," in which he gives his interpretation of these results. The last section, "Further Evolution," does not correspond even roughly to a section of a conventional article, but crowds in some of the ideas from the early draft, relating all this back to his broader claims. The form still looks like a personal essay embodying the researcher's thought, rather than a research report embodying the discipline's criteria for judgment.

Although the organization of Bloch's first submitted version suggests big ideas, his tone is as cautious as he can make it. "A panoramic view of evolution offers clues that can serve as a guide in ordering the early stages." The tentativeness of the diction balances the enormous claim; he finds "clues" and a "guide," not a demonstration. When he describes the model in his introduction, the verbs are almost all conditional ("*could* provide a configuration") and the claims tentative ("*is envisioned* as a hairpin structure"). His characteristic method of argument, here and elsewhere, is to survey a broad question, suggest possible answers, and argue against the alternatives until only his own view is left. He tries to give the impression of a balanced approach, but the responses of the referees indicate that he does not successfully avoid giving the impression that he has a prior commitment to one interpretation, that of common origins for tRNA and rRNA.

The mixture of boldness and caution in Bloch's tone is apparent in his presentation of what he told me was "gratuitous but suggestive evidence," a ratio, which he saves for last, involving the information content possible with the number of RNA nucleotides. As he puts it in the earlier version, "This is a tantalizing bit of numerology that evokes no ready explanation from current views of RNA functions or relationships." On the one hand, he is claiming to introduce a new view of the evolution of life; on the other he injects his characteristically self-mocking tone with "tantalizing" and "numerology."[8] A conclusion that Bloch added in the version after this one can serve as an example of the style of much of his writing. "The scattered homologies are likened to the shards with which the archaeologist reconstructs pottery of ancient civilizations." The awkward sentence structure shows how hard it is to work this simile, which Bloch used often in his oral

8. For a comment on this term, see M. Gossler, "Numerology," *Nature* 306 (1983): 530.

presentations of his work, into the passive constructions of the scientific article. We could treat this mixed tone, like the exploratory organization and metaphorical conclusion, as a tactical error on Bloch's part, perhaps as evidence of inexperience. But since we know he could and did write straightforward research reports earlier in his career, it seems reasonable to take these departures from form as assertions that his claim is important enough to justify some background for the argument, some speculation in the conclusion, and some personal style in the presentation.

Whereas Bloch's departures from the form of the research report can be seen most clearly in the structure of his manuscript, Crews's departures from the form of the review article are largely a matter of tone and emphasis. And he is aware of the effect of these departures, again because of his immersion in a network that gives him responses before he submits a manuscript. As I have suggested, Crews was guided in his reframing of the draft for publication by the marginal comments, sometimes quite acidic, of a number of his colleagues. For instance, one reader points out, "It takes a long time (many paragraphs) before you get to the *new concepts*," and responds to an "indirectness" in the argument by proposing "a different strategy of organization" which he describes clearly, and which Crews adopts. Another reader raises potentially troublesome questions about the physiology of a particular species, giving the kind of detailed argument one seldom sees in a referee's report. A graduate student working at another lab where Crews has contacts compiles a three-page list of ambiguous phrasing and terminology. In each case, the reader defines part of the potential response of the zoological and endocrinological communities, before Crews submits the manuscript for judgment. All these different styles of handwriting in the margins of various drafts are the visible sign of the invisible college.

Despite these suggestions, Crews's first submitted version shows some tension between what he wants to say and the review article form in which he must say it. We can see these tensions by comparing the tone of some passages of this manuscript with similar passages in an earlier review article that Crews published in *Science* in 1975, when he was a postdoctoral student and was perhaps more cautious (Appendix 2). A review article typically summarizes the recent work of a research program, drawing on a broad survey of the literature, tactfully and impersonally presented. So the 1975 article has the unthreatening textbook-like title, "Psychobiology of Reptilian Reproduction." But the title of Crews's 1983 manuscript says he will give "New Concepts in Behavioral Neuroendocrinology," challenging the work of this research program. The earlier article begins with what might be considered the stereotypical opening sentence of a scientific review article: "The interaction of behavioral, endocrinological, and environmental factors regulating reproduction has been the subject of intensive investigation in recent years." The diction of the first sentence of the new article is provocative and even combative: "Much of the information on the causal mechanisms of vertebrate reproductive behavior has been gathered on *highly inbred stocks* of rodents and birds living under *artificial conditions*. . . . Some of the organismal level concepts that have emerged are *overly narrow* and *sometimes unrealistic*" (emphasis added). I assumed that the phrases I have emphasized would be red flags to other naturalists: he is saying that they are studying something unnatural. He uses a vocabulary with contrasting connotations to describe his own work; he proposes to investigate "species diversity under naturalistic conditions" and quotes comparative biologists who say such an approach leads to "new insights" and "new paradigms of thought." He is particularly bold in attacking the most commonly studied species as well as the most commonly held ideas; psychologists have money, time, prestige, and egos invested in their laboratory animals, and might respond more fiercely to attacks on their mice than to attacks on their minds.

A similar sharpness of tone is apparent in a comparison of the conclusion of the 1983 manuscript with that of the 1975 article. The earlier article ends with a concession to the competing research program in a subordinate clause and a conventional reference to the continuing advances of the field: "Thus, while the utilization of inbred species contributes greatly to our understanding of the factors regulating reproduction, the integration of these factors can only be appreciated fully in an ecological context where the adaptive significance of such interactions become apparent." The 1983 article ends with a statement of a similar idea, but frames it in terms of a call for more work on the whiptail lizard (*his* species), so that "it becomes possible to apply evolutionary theory to gain insight into the evolution of psychoneuroendocrine mechanisms." The earlier article stresses uses of *our* knowledge, whereas the later manuscript suggests that a whole new approach has been overlooked by most workers in the subspecialty.

From a rhetorical point of view, we might argue that the tone of these sentences is a strategic mistake, a miscalculation of his audience. But seeing the article in terms of a negotiation, we can see his tone as an assertion of the value of his knowledge claim. He is saying that this article differs from the views of most neuroendocrinologists, but it is still important enough for the front section of *Science*. A more cautious article presented as a review of current knowledge, with a title like "More on the Psychobiology of Reptilian Reproduction," would be more appropriate in tone, but less important to nonherpetologists, and thus less appropriate for *Science.*

Many of the reviewers' comments are concerned with the departures from standard organization and tone I have described. For instance, all Bloch's reviewers comment on the length of his manuscript, even though he revised it, after the first rejection, to fall just within *Nature*'s word limits, pointing out that it is "about 2,930 words" (*Nature*'s limit is 3,000). As with most academic journals in which space is at a premium, appropriate length is determined, not by the limits given in the "Instructions to Authors," but by the importance granted to the article's claim; these reviewers do not think Bloch has earned 3,000 words yet. The most telling criticism of Bloch's style comes from his most enthusiastic referee at the journal that finally publishes his paper, a reader who seems to worry that Bloch's persona will endanger the reception of his work. "If the author is to have his observations seriously evaluated by others in the field, it is important that he not present himself as being overly speculative. Discussions of 'shards' and extremely speculative ideas such as Figure 5 [his original model] and those beginning at the bottom of page 7 [interspecies comparisons that form the basis for his current work] will not improve the author's chances of being taken seriously at this stage and would best be removed." This response shows that Bloch is perceived as a newcomer to the field, whose use of personal metaphors, asides, and "notions" is inappropriate "at this stage," and who may need guidance on his presentation. Perhaps a more personal and expansive style is permitted to those whose work has already been recognized.

Some of the rather vague criticisms from reviewers of the form and style of Crews's paper also seem to be directed at his departures from convention, in this case the format of the review article. For instance the more favorable referee of the first version says "the manuscript is not well-written." It is always hard to know exactly what this sort of comment means, but if we read on we find the more definite criticism that "a review paper of this nature which has pretensions to generalization should not be based on a preliminary review of the literature!" The meaning of "preliminary" here is relative; the 1983 article has more references than the review article published by the same journal in 1975, and probably has, already, more than *Science* wants to print. The problem is, perhaps, that a review article must not be so much a review of one's own work, in which the work of others serves mainly as a background; the reviewer could be objecting, not to the number of citations, but to the emphasis implied in the organization. A favorable reviewer of the second manuscript submitted to *Science* shows, as we saw in Bloch's case, that criticisms of form—especially length—can sometimes be interpreted as attempts to redefine the claim. "A short paper will be read more often—a point briefly made is often the point well made." This may be good advice for any academic author, but the specific passages the reviewer would like to cut suggest the reviewer is more concerned Crews will alienate readers than that he will bore them. The garter snake sections can be deleted because "anyone interested in reproductive biology must have noticed the article by Crews in *Scientific American* a few months back." This change, like the comment on the "preliminary" review, can be read as an insistence that he move the emphasis from his own work. The reviewer also suggests that the whiptail lizard sections should be cut because they are controversial. "Female mating in the wild has never been observed (judging from a heated discussion by *Cnemidophorus* workers after a seminar by Crews at a recent ASZ symposium.) . . . Personally, I think Crews is on shaky ground here, and there would be great danger of a hostile reaction to an otherwise important contribution." The reviewer does not attack the *Cnemidophorus* work directly, but can rely on this vague consensus. The work is inappropriate in a review article that claims to represent the work of the specialty, not because it is wrong, but because it is the author's own claim and has not yet been accepted by others working on the same animals.

Negotiating Form and Style

The changes the authors make in various revisions in response to these reviews show they take these apparently superficial matters of organization and style as issues affecting the status of their claims: they make most of the changes suggested, but reluctantly. Bloch described his revision of the article for *Science*, after he read the *Nature* reviews, as "cutting some of the speculation and adding some new data." This he certainly did, extending his list of matches and including his recent reading in the reference list. He also changed his self-presentation radically, becoming the judge of, rather than the advocate for, the claim for the common origins of tRNA and rRNA. The title, "An Argument for a Common Evolutionary Origin of tRNAs and rRNAs" becomes "tRNA-rRNA Sequence Homologies: Evidence for a

Common Evolutionary Origin?" The new title puts the data first, changes an *argument* to *evidence,* and strikes a note of skepticism with the question mark.

The structure of Bloch's new version moves much closer to the structure of a conventional research report. Comparison of the first version submitted to *Nature* and the revised version submitted to *Science* shows a reorganization along inductive rather than deductive lines, moving toward the conclusion of common origins rather than from this assumption. The first four pages of the review of theory are cut, as are the last two pages on prospects for "further evolution." The article now begins with the homologies, lists some of them, and gives his methods. The formula for determining the significance of these homologies, his link with a recognizable line of previous research on sequences, is now on page 2 instead of page 6. In general the exposition is tightened, introductory sentences are added, a few asides cut, and some sections are moved from "Results" to "Methods," where they flesh out that previously rather skimpy section. Bloch split the section on whether the homologies result from convergence to take into account both convergence and function, showing a new refinement in his argument and recognition of a body of literature on sequences. The section on convergence, his refutation of a counter-interpretation, is shorter, and ends cautiously saying that a larger data base is needed. The model that was first is now last; it occupies only a paragraph, and comes with no elaborate explanation of the conditions it satisfies. The article is six pages long instead of twelve, with ten notes instead of sixty (*Science* discourages "exhaustive" reference lists). But only notes indicating the sources of sequences are cut; all the substantive references to related work by others are retained.

Bloch's revisions for the *Journal of Molecular Evolution* continue the reorganization into more conventional format, with an emphasis on the data, and into a less personal and less assertive style. The introduction emphasizing the significance of his findings is cut, and a short summary of his method is put in its place. In the first sentence, where before rRNAs were "peppered with stretches" homologous to tRNAs, now they "were found to contain stretches. . . ." In a gesture toward consideration of both sides of the data, suggested by a referee at *Science,* he adds a new table showing the tRNAs that *don't* have homologies. The conventional heading "Discussion" replaces the earlier, less formal heading, "Why the Homologies?" He finds more arguments against the possibility of coincidence or horizontal transmission and supporting the concept of a multifunctional molecule. He makes the tone even more cautious: "We propose that" becomes "one interpretation would be that. . . ." Finally, the model is deleted entirely from the text and relegated to the caption of figure 5, and the sentence on how the model led to the finding of the homologies, the last relic of the narrative of his thought that Bloch gave in his first version, is deleted.

After the review at the *Journal of Molecular Evolution,* Bloch makes nearly all the changes the reviewers and editor suggest. He adds a numerical example, some figures on the quality of the matches, an example of his calculation of the possibility of coincidence, and references to possible DNA-level transfections. The tone becomes still more cautious; an "acceptable" level for excluding homologies as coincidental has now become a "provisionally acceptable" level, and where he had said that these coincidental matches "will be revealed" by further comparisons, now he says they "should be revealed." He admits a possible weak point of his method of argument, that "the evidence so far has supported homology only by eliminating or weakening arguments favoring alternative explanations." And finally he deletes figure 5, which was criticized by the reviewers, and with it all trace of his model.

In addition to the changes suggested by reviewers, Bloch makes an apparently minor formal change suggested by the editor that is relevant to the position of the article in the literature, the same change from *homology* to *matching sequence* that we saw in the proposal. The editor had commented on Bloch's use of the term "homology," a complex term that usually means a *common sequence* in molecular biology, but means a feature resulting from *common origins* in evolutionary biology.[9] The editor said he had long had a policy of trying to keep the word univocal, and argued that use of the word in its more general sense marks an unnecessary division within the discipline: "molecular biologists have to be biologists too." Bloch was happy to agree and change his use of the term; otherwise he would be begging the question in arguing that the homologies showed common origins. However, he also adds a note saying, "Their distributions suggest . . . that they represent true homologies."

The concluding metaphor of the shards is gone, alas. Instead Bloch ends the published paper with another metaphor, that of "filling in the map." But this, he tells me, refers to "a phrase used back in '55 by Benzer, describing filling in the genetic (linkage) map with mutants,"

9. Mayr, *Growth of Biological Thought,* p. 465; M. Norell, "Homology Defined," *Nature* 306 (1983): 530.

so it is an allusion to a tradition in the field, not an assertion of a personal style. We may, however, see a personal style in another sort of figure added by Bloch to the final version of the paper, a diagram related to his current work on species comparisons. The version in print is full of tables and graphs; as a colleague said, "every day he thinks of some new way to illustrate it." His graphic figures may in some way replace the figures of speech he has had to cut; in fact, his figure 2a, showing where various matching sequences are located on a conventional diagram of tRNA, is the equivalent of his shards metaphor. Both kinds of figures provide visual images that represent selected features of highly complex data. In these terms, Bloch gradually changes his figures to the kind more acceptable to the *Journal of Molecular Evolution*.

Crews's revisions of the first version submitted to *Science* also show some concessions to the views of the referees and the conventional form, with its implied placement of the claim in the context of the literature, in order to get his claim in print. In his first revision for *Science*, the accounts of his own studies are shortened and subordinated to the work of others. This version is more readable: digressions are deleted, especially near the beginning, transitions are added, some supporting but complicating details are moved to the notes (he now has eleven explanatory notes instead of two), and a concluding restatement of the argument replaces the appendix-like anticlimax of the earlier draft. He says in his letter asking for a second review that these changes make this version more "straightforward," but these changes affect the persona of the article as well as its readability, for the article now makes a sharper claim and makes fewer demands on the reader.

I can draw no line of demarcation dividing the changes Crews is willing to make from those he is not. But in general he is acutely aware of how his tone defines his relation to the work of others, and he is willing to change this tone wherever necessary. He is unwilling to modify his inferences from his evidence, preferring even to cut sections and use them in other articles rather than compromise his argument. The change in tone at this stage is suggested by the title; the assertive "New Concepts in Behavioral Endocrinology" becomes the descriptive "Functional Associations in Behavioral Endocrinology: Gamete Production, Sex Hormone Secretion, and Mating Behavior." The provocative opening remark about other researchers' "highly inbred stocks of rodents and birds" becomes a milder comment on "laboratory and domestic species." Where before, in the summary attacked by one reviewer, he said "this survey makes several points", now it "raises several questions." He adds a cautious note in saying that the lack of dependence of mating behavior on hormones "may be more common in vertebrates" than previously thought. Other changes in tone are apparent in his changing "my laboratory has been investigating" to "the most thoroughly investigated species . . ." and in his phrasing of an assertion in the form, "it is important to restate the obvious." He responds to a reviewer's criticism of his "really very simple model" by pointing out that "the four reproductive tactics . . . represent extremes." He is careful to incorporate the "existing body of knowledge" referred to by the other reviewer, reminding the reader that many species *do* follow the conventionally accepted pattern. An example of his avoidance of confrontations (and witty understatement) is his mention of his most controversial point, the relevance of all this to humans, only in a note. By softening the confrontational tone of the earlier version, Crews includes his readers on his side of the argument, whether they belong there or not.

Although Crews backs off in matters of tone in this resubmission to *Science,* he mounts a counterattack in the form of his argument, adding a great deal of material. First he establishes the paradigmatic status of the concept he is attacking in a new transition: "the concept . . . has persisted despite an increasing number of studies revealing variations to this rule." Here he adds a number of counterexamples and then asserts, cautiously, that "It is possible that the rule . . . may be due to a bias in the species most studied." Since his associated/dissociated dichotomy was considered too simple, he adds more examples to develop it in detail. The brief comment that was called "a quantum leap from faulty premises" is expanded into four paragraphs. Another comment that had been criticized, on explosive or opportunistic breeders, is moved from the beginning, where it seemed an aside, to the end, where it is introduced as "a classic example," well known to all.

The article can no longer be called "a preliminary review of the literature," and if it is to become, in the words of the reviewer, "a straw man," it is well stuffed. The revision is only one page longer, but whereas the earlier version had 57 references, the new one has 195, far more than is usual in a *Science* review article. The considerable changes show again how Crews's place in a research network of zoologists, psychologists, and endocrinologists allows him to respond to critics in his revision. The earlier version had listed twelve readers, mostly colleagues in the zoology department; the second lists thirty-one, mostly from other departments and schools. And this list includes only the actual readers, not those who raised questions or made criticisms and suggestions after the many lectures he gave while he was revising, or those who talked to him in the halls, on the phone, and after work at a Mexican restaurant. Any writer can cut the parts of an article criticized by reviewers, but perhaps only a writer who argues his claim every day can rebuild the article on broader foundations of evidence in a period of a month.

This remarkable flexibility continues in the five major revisions Crews made between the *Science* version and the one for *Nature*. At first, like most authors, he tried to make a minimum number of changes, using his word processor to change all occurrences of *behavior* to *behaviour* for the British journal, cutting the subtitle, which exceeded *Nature*'s limit of eighty characters, and adapting his references to its style sheet. But in later drafts, after many more readings by colleagues, he made what he considers "wholesale cuts." What finally emerged is a very concentrated version of six and a half pages (of which three and a half are devoted to a figure, a table, acknowledgments, and references), with the new, catchy, headline-style title, "Gamete Production, Sex Hormone Secretion, and Mating Behaviour Uncoupled." The introduction is gone, and the paper begins immediately with the argument. Following a favorable reviewer's suggestion, almost all examples are relegated to the table, only two sentences are left on Crews's garter snakes, and his controversial *Cnemidophorus* studies are deleted entirely. One important sentence is added, making the current view that he is attacking seem one-sided: "Indeed, all of the data supporting this paradigm have been obtained from species in which both sexes exhibit an associated reproductive tactic." Now there are just fifty-two notes; significantly, only five of them refer to work done in his lab, and the first of these is carefully placed far down the reference list. The evolutionary argument the *Science* reviewer had called "a clever and reasonable hypothesis" is now apparent only to the reader who compares a statement on the second page (saying that the old view had supported phylogenetic conservatism of these relations) to the last sentence ("The possibility that similarities in the mating behavior in different vertebrate species [are] the result of convergent, rather than divergent, evolution, adds another perpective to our understanding"). As the form of the article has approached the conventional format, and the tone has become more cautious, the article has changed subtly from an attack on a paradigm by one scientist to an outline of the logical implications from the collective work of all the researchers in the field.

How do "*J Mol Evol* (1983) 19:420–28" and "*Hormones and Behavior* (1984) 18:22–28" differ from the authors' first manuscripts, "The Evolution of Evolution" and "New Concepts in Behavioral Endocrinology"? The claims in the published versions are at a lower level of the hierarchy, Bloch claiming only that matching sequences *may* indicate common origins, Crews claiming only that his comparisons show that gamete production, sex hormone secretion, and mating behavior may be dissociated in some species. In Latour and Woolgar's terms, they have had to add modalities and move their claims away from fact-like status. In Pinch's terms, the authors, in this evidential context, have to settle for claims of somewhat less externality than those they had first proposed. They have to leave out their models, and this could be a loss for them, because whatever words have been excluded at this point, as the article goes into print, cannot be part of the authors' claims. So if, for instance, molecular biologists not only accept Bloch's claim of common origins for these two molecules, but follow this claim to something like his model as well, this article would give him no way to assert his priority. (For this reason, he described the model in an abstract published separately.) We see this limitation of claim as well in the more conventional personae and forms the authors use in the published versions. These are, as one might have guessed, not so much fun to read as the earlier drafts, and not so clear to a nonspecialist, since they are highly compressed, are allusive in their references, and give none of the background or history of the claims.

Perhaps the most serious change in the articles, in practical terms, is that they now reach much more limited audiences than those the authors had hoped to address when they submitted their manuscripts to *Nature* and *Science*. This means not only that the articles miss whatever prestige an article acquires by appearing in those journals, but also that they are less likely to be seen, in Bloch's case, by the molecular biologists doing sequencing, and in Crews's case, by the wide range of zoologists. These are the researchers who, if they reoriented their research programs to pursue these new interpretations of published data, might provide more data to strengthen these claims: more sequences to check for matching tRNAs and rRNAs, or more animals for which the patterns of hormone levels, gamete production, and mating behavior are reliably known. But we should remember that Bloch and Crews are asking for a great deal. As with grant proposals, the selection process serves a social function. If we wanted to explain all aspects of the scientific community in functional terms, we could see in the relegation of their articles to more specialized journals an example of how the publication process works, protecting these researchers in other fields from just this kind of claim from outside their own research programs, and thereby preventing the capricious redirection of goals, the proliferation of research programs, and the scattering of resources.

The process of publication of a claim does not stop with the acceptance of one article; both writers have other outlets for their ideas, at other levels of the hierarchy of journals. Here again we see a sharp difference between Bloch's opportunities and those of Crews, with their differing positions in their fields. Bloch tirelessly presented his papers on RNA sequence matches at conferences, and argued his views with important speakers visiting his department. In one poster session at a huge cell biology convention, he would repeat to anyone who was interested his whole case for common origins, drawing the listener along from figure to figure. If the listener seemed interested, Bloch might go on to his larger ideas about surrogation. Thus, in this forum he could have his own form and choose his own level of claims, according to the responses of the individuals who made up his audience. But his audience on this occasion consisted largely of friends and students, nearly all still working in his old field, and a few passers-by, perhaps attracted first by his lively illustrated bulletin board, many of them apparently graduate students with time for an intriguing, if odd, idea. Bloch put a great deal of preparation and energy into these presentations, and was happy with the chance to persuade anyone, but it seemed unlikely that he would persuade in this way the powerful molecular biologists whose interest he needed.

Bloch found another outlet for his model in a very short version of a paper delivered at a European conference on the Origins of Life, the proceedings of which were then published. In this unrefereed outlet he was freer to speculate, as the less cautious title suggests: "tRNA-rRNA Sequence Homologies: A Model for the Origin of a Common Ancestral Molecule, and Prospects for Its Reconstruction." This gave him a citation he could use in proposals and in other manuscripts to refer to his model, and a priority claim for the idea of primitive multifunctional RNA, should the idea be widely accepted. But even he found it rather too compressed to be easy reading. And he discounted the authority this publication would have for the experimentalists he needs to reach; with some praise and self-irony he called the origin-of-life people (among whom he counted himself), "a bunch of nuts." He said that later when a more detailed paper that included the model was published, the early paper would be superseded.

So Bloch continued to try to find outlets for the parts of his work cut from the published article and for the data and theoretical refinements that emerged after the final version of the *JME* manuscript. He submitted to *JME* a second article arguing that interspecies comparisons would show evolutionary convergence, and a third article with the details of the model, but these were both rejected. Then his collaboration with the physicist at the Center for Statistical Mechanics, Apolinario Nazarea, led to a manuscript in which "second order spectral analysis is used to depict rigorously and to characterize principal periodicities in the positions of conserved sequences common to tRNAs and rRNAs." Note that the last phrase, "conserved sequences common to tRNAs and rRNAs," takes as proven, not only the existence of matching sequences, but also the explanation that they are due to common origins. But, as the earlier reviewer's comment about publishing the data first suggests, articles like this had to wait until his earlier findings were known enough to serve as the basis for new problems. The publication of the *JME* article was helpful, but it did not become a breakthrough that would open further outlets for publication; it was not immediately cited and did not immediately become a part of the literature on molecular evolution.

For Crews the question is not so much whether an article can be published as where. Even though, as we saw, he cut out the first half of his manuscript before sending it out, and finally published an article six and a half pages long, he has been able to publish most of what he wrote. The material on his own work, which he cut to place more emphasis on the field as a whole, appears in two articles in an issue of *BioScience,* a glossy but rather serious popular biology journal. He was guest editor of this issue, chose seven articles (including his own) on similar comparative research, and used the forum to make his polemical methodological point about the importance of studying atypical species. Crews has also edited a book gathering together studies that show alternative reproductive tactics in a wide variety of species in all the major categories of animals, *Psychobiology of Reproductive Behavior: An Evolutionary Perspective.* This, too, is a kind of outlet available only to researchers who are already well established in a network of other researchers, to whom they can turn for work parallel to their own.

Though the theoretical implications of Crews's claim were cut from the *Hormones and Behavior* article, or at least well hidden, he was able to present them undiluted in a paper for an unrefereed but not unprestigious forum, an invitational symposium at the Kinsey Institute (1985). For this audience of physicians, psychotherapists, and other researchers interested in sexuality, an audience that did not need to determine the status of his claims or place him in the literature of neuroendocrinological research, he could be as assertive as he was in earlier drafts. The argument had become more cautious since then, and is supported by all the additional data gathered during his revisions. But the tone, even in the abstract, is like the tone of "New Concepts in Behavioral Endocrinology": "The great diversity in reproductive tactics . . . has been unappreciated by behavioral endocrinology," and "the deterministic paradigms in behavioral endocrinology are overly narrow." Where in the *Hormones and Behavior* article the evolutionary ideas were held until the end, here they are emphasized from the beginning. From the out-takes of the *Science* version he gets a sentence on the evolution of regulatory mechanisms, lists of exceptions to the paradigm and of animals with associated or dissociated patterns, and descriptions of his own work. Two pages on the development and evolution of functional associations at the end of the *Science* submission, which had been the focus of criticism from the more hostile referees, are here expanded into six pages. The added pages make explicit the way his claim applies to other levels of the study of reproduction, so the place of the *Cnemidophorus* in this program is now clearer.

A particularly telling difference between the Kinsey talk, for a general scientific audience, and the *Hormones and Behavior* paper, for an audience of neuroendocrinologists, is in Crews's use of the scientific literature. He begins the Kinsey talk with a motto (a practice common in reviews by elder statesmen, but not usual in a scientific paper) from an article dating to 1946, and he refers prominently in the introduction to insights from masters in the comparative field, often from texts twenty to forty years old. The quotations seem to be a part of the persona he is developing here; on the one hand he is an outspoken dissenter from the rigid paradigm of neuroendocrinology, but on the other hand he is the inheritor of a rich tradition of comparative work. Although such self-presentation is not encouraged by the review article format, it is appropriate in this oral presentation to an audience of nonbiologists, for whom he must represent biology (he is the only biologist there) and also present something lively, new, and relevant to their own work with humans.

Finally, almost all of Crews's first article appeared in print. But it appeared in five separate texts, for four separate audiences, and was inserted into the structure of scientific facts in four different ways. For Crews as for Bloch, the way his text enters the literature is crucial in determining the eventual status of its claim. Whether his claim or Bloch's becomes a fact depends on how the articles are used by other researchers. But the form of the claims has been set; even if we don't know what the response of the research community will be, we know exactly what they will be responding to. What is not printed cannot be cited.

Though it is still too early to judge the fate of Bloch's model of molecular evolution and Crews's model of diversity in reproductive systems, the first responses to the articles can be taken as an ironic postscript to my story, and as a reminder that a bumpy ride from the referees and publication in a specialized journal do not necessarily mean that an article will remain in obscurity. Two months after Crews's article appeared in *Hormones and Behavior,* an editor of *Science,* interested in the issues it raised, called and wrote to Crews to offer an official invitation to write a longer review article on this topic. Even after all his experience revising the *Hormones and Behavior* article, Crews rewrote his submission drastically several times, and had his coauthor Mike Moore (with whom he had developed many of the ideas) rewrite it again after that, before it was submitted, reviewed (by two favorable reviewers this time), and accepted. When it appeared in January 1986 as "Evolution of Mechanisms Controlling Mating Behavior," he had finally published his main claim in the form in which he wanted it, in the journal to which he had originally sent it.

Bloch had a similar ironic turn of fortune. When he began his collaboration with Nazarea, he was skeptical about submitting their statistical analysis to *PNAS;* Bloch said in a letter that sending it to such a prestigious general science journal was "his idea, not mine." Rather to his surprise, it was accepted by this prestigious general science journal. The model appeared in *BioSystems,* which had invited Bloch to submit it after the origins of life conference. And soon afterward these two articles were discussed in a page-long news article in *Science* by Roger Lewin. A passage from the article shows how Bloch's work can be placed in the literature so that it is news, and disagreements are presented as controversy, rather than as rejection.

> Although their conclusion is not universally accepted, Bloch and his colleagues consider that there is sufficient reason to argue that the sequence similarities between tRNA and rRNA between the species reflect a common origin, not a recent convergence, and is therefore homologous [cites *BioSystems* article]. A similar, but much less detailed, suggestion based on comparisons of a tRNA and a small (5S) rRNA was made more than 10 years ago by James Lacey and his colleagues at the University of Alabama (2), but it was not extended to the larger rRNA's that are the basis of the Austin study.

The way Lewin refers to "the Austin group" and "Bloch and his colleagues" must have been a satisfying recognition that he did at last have some people to work with, "someone to bounce ideas off." Bloch was preparing further articles when he died in the autumn of 1986.

The turnaround in the fortunes of both claims, so soon after publication, reminds us that, though historians can locate classic papers in retrospect, the success or failure of a claim seldom hinges on one article. Acceptance or rejection usually comes over the course of many articles, and last year's wild speculation may become this year's plausible hypothesis and next year's basic assumption.

Conclusion

What level of claim can I persuade the readers of this book to accept? At the lowest level, I am saying that scientists sometimes revise their manuscripts considerably to get them published. To support this claim, I need only show you the stacks of manuscripts. This claim, though on a very low level of externality, is significant in some evidential contexts, for instance, the context of technical writing teachers trying to convince their students of the value of rewriting assignments, or perhaps the context of scientists displaying to nonscientists the work that goes into a seven-page article. But for the audience I hope to address here, that is arguing on the level of Pinch's "splodges on a graph," the level of uninterpreted data. I can put the claim on a higher level of externality, to continue using Pinch's terms, by using these manuscripts to show that a scientific claim is socially constructed. But this claim, though it is in terms familiar to sociologists and historians of science, tells them nothing new. A more specific claim, that is likelier to tell the readers something new, is that the comments on and revisions of these manuscripts show one of the ways in which claims are socially constructed, that is, through the negotiation of the form of the article and thus the status of the claim. It seems to me that this claim may have relevance in two different evidential contexts, telling us about science or telling us about texts. In one context, these cases suggest that the process of writing and revision of articles has an important consensus-building function. We have seen how this process maintains the homogeneity of the scientific literature. We have also seen how it shapes the research itself, Bloch, for instance, putting more and more emphasis on his data. In another context, that of literary criticism, these cases show the relations between texts, within the genre of the scientific article. The question in this context is not how reality is transformed in texts, but how it is made by texts.

Like Bloch and Crews, I need more data to support my claims, so I address in this book an audience interested in scientific texts, though perhaps interested in them in other evidential contexts. I particularly need data from other disciplines and earlier and later stages of the publication process, about how persuasion is planned before a draft is written and how an article is read after it is published. From cases in other disciplines, it would seem generally true that a number of claims are possible from one line of research, and that disagreements about the status of claims do tend to focus on matters of appropriateness to the journal, organization and length, persona, and use of the literature. But I haven't yet seen enough descriptions of the publication process in the literature to know how far this description is useful. I see from Pinch's cases that negotiation in physics and biology are rather different; the biological arguments seem to involve the usefulness of alternative concepts for organizing large bodies of data that were collected for other reasons rather than the sort of crucial experiments that characterize the history of physics. For instance, it seems to be an acceptable response to Crews's argument from the mating of garter snakes to say, "that's just one species," whereas it is not an acceptable response to an argument from the perihelion advance of Mercury to say, "that's just one planet." We may find other characteristic differences between disciplines in the process of negotiating a claim and a published text.

The earlier and later stages of the process of writing a scientific article seem to be particularly appropriate for ethnomethodological and ethnographic approaches.[10] Historians of course, do not have access to the daily phenomena of a lab or to the immediate responses of readers, but they do have access to a wealth of written texts that may be more revealing than any direct observation.[11] The tendency of contemporary history of science to massively documented biographies of key individuals, although it may lead away from crucial sociological issues, is likely to yield insights for students of texts (the

10. Lynch's *Art and Artifact* and Garfinkel, Lynch, and Livingston's study of a tape of astronomers at work are particularly good examples of detailed documentary study of events and interactions preceding writing. See H. Garfinkel, M. Lynch, and E. Livingston, "The Work of Discovering Science Construed with Materials from the Optically Discovered Pulsar," *Philosophy of the Social Sciences* 11 (1981): 131–58.

11. Martin Rudwick's *The Great Devonian Controversy* is a classic study that may remain, for quite a time, the fullest possible use of texts as documents; Frederick Holmes, "Lavoisier and Krebs," has traced the development of a text by Lavoisier, and Charles Bazerman has begun a similar study of Newton's *Optics* in *Shaping Written Knowledge*.

enormous literature on Darwin is an example). For the later stages of a knowledge claim, after publication, there have been some interesting studies of readers' interpretations.[12] But to see these stages, one has to go beyond case studies of individual scientists and laboratories to the core set (Harry Collins's term), the group of researchers concerned with one research issue. The fate of a claim is not decided when it is published, even when it is published in *Nature* or *Science;* it depends on who reads it, how it is read, and how it is used.

12. Studies of readers' interpretations include Amman and Knorr-Cetina's forthcoming study of conversational responses in the representation of a gel in a molecular genetics article, Gilbert and Mulkay's interview material on the evaluation of experiments ("Experiments Are the Key"), and Charles Bazerman's chapter on physicists' reading in *Shaping Written Knowledge,* "Physicists Reading Physics."

Chapter Four

The Cnemidophorus *File: Narrative, Interpretation, and Irony in a Scientific Controversy*

A nonbiologist might expect, after reading the exchanges between reviewers, authors, and editors quoted in chapter 3, that the publication of Bloch's and Crews's articles, and the further publicity given to their claims in *Science,* would be followed by the sort of heated controversy familiar to readers of the *New York Review of Books* or *Critical Inquiry.* But this sort of back and forth exchange in print is not common in the scientific literature. The usual method of dealing with research claims one thinks are wrong is to ignore them; if they are not picked up by anyone, they will disappear into the morass of scientific publications. Citation analysts have often noted that negative citations are rare; the lack of any citation is a much more effective way of dismissing a claim.

Though printed exchanges are rare, controversies are quite common in science, probably much more common that the nonscientist imagines. Sometimes they concern priorities, when the research is perceived as a race toward a clearly defined goal. But more often they concern the definition of the goal of research, the conceptual framework within which work is to continue. These controversies are pursued at conferences, in phone conversations, in letters, in referees' reports and in implicit comments in articles. Just as the usually unnoticed dynamics of article reviews are clearest in the rare cases (like those in chapter 3) in which an article is repeatedly revised and resubmitted, the informal and implicit exchanges of scientific controversies are clearest in the relatively rare occasions when they emerge explicitly in texts. That is why controversies have been so important to sociologists and historians, like those described in chapter 1, who want to open up the black boxes of science, to show the processes of construction of what we take as facts.[1]

I have a brown manila folder that contains reprints of articles by three groups of biologists, sent to me by David Crews with a note reading, "Controversy and how I have handled it." Taken separately the articles look unremarkable enough: reports of results in various studies of the behavior of lizards. But taken together they tell a story, at least to the biologist who labels them a "controversy." There is a crucial rhetorical difference between the strategies in such cases and those considered in chapter 3. The immediate audience in a controversy is not the editor or the referee who controls access to a journal, but the broader group of researchers working on whatever is defined as the problem, especially those researchers not already committed to one view or the other of the work at issue. They raise questions about how a writer can open up discussion of a scientific disagreement, how one text relates to another, how disagreement is finally resolved or ended, and how the fact that results is codified.

One way I see of approaching the questions raised by this file is by looking at the articles as showing the contruction, interpretation, and reinterpretation of narratives. They are not just evidence of a controversy; they are presentations of that controversy, of where it comes from, what it is about, and how it should end. (Even Crews's short note labeling these texts a controversy, and his selection of these texts, might be disputed by the other participants.) What interests me about this process is what happens before one story wins out and becomes *the* story.

By *narrative,* I mean the selection and sequencing of events so that they have a subject, they form a coherent whole with a beginning and an end, and they have a meaning that is conveyed by the sequence as a whole. If this seems an odd activity for scientists, it may be because we associated narrative with storytelling, fictions, and falsehoods. Even some critics who give narrative a more general definition insist on a distinction between narrative and the information that makes up scientific knowledge. For instance, Jean-François Lyotard refers to "the preeminence of the narrative form in the formulation of traditional knowledge" when distinguishing this traditional knowledge

1. The most detailed, and best written, of accounts of controversy is Martin Rudwick's *The Great Devonian Controversy.* See also Andrew Pickering, *Constructing Quarks;* Harry Collins's studies in *Changing Order;* Harry Collins and Trevor Pinch, *Frames of Meaning: The Social Construction of Extraordinary Science* (London: Routledge and Kegan Paul, 1982); Steven Shapin, "The Politics of Observation," and the introduction for general readers in Latour's *Science in Action,* which cites a number of other studies.

from knowledge in "the scientific age".[2] Narrative statements are taken as true within a context and a form; scientific claims are supposed to be true regardless of context, precisely because they have been stripped of context, of actors and processes.

This distinction between narrative and scientific argument has long been respected by literary critics; we saw in chapter 1 that when Dwight Culler treats the *Origin of Species* as a narrative constructed by Darwin, he is no longer treating it as science, and when Walter Cannon treats the *Origin of Species* as science, he dismisses the form as a distraction. But recently there have been a number of studies of narrative in scientific rhetoric, by literary critics, anthropologists, and sociologists of science. These approaches are quite different from each other, but it should be pointed out that none of them is a debunking exercise showing something unscientific in storytelling.[3] They are not showing the scientists doing something scientists are not supposed to do, because they are not assuming that there is a kind of knowledge stripped of narratives, to which scientists are supposed to restrict themselves.

The scientists themselves might see the controversy as working through arguments, rather than narratives. Arguments are supposed to work by reference to evidence and to the prescriptions of inductive reason, that is, by standards external to the discourse. But the power of narrative is based on its form, and this formal power is at work much of the time in this controversy. The biologists do bring in evidence, but it is effective, or isn't, because of the way they make the whole story fit together so that it has meaning; change one part and the whole meaning changes. We shall see a frequent tension between the author's assertions that the texts are arguments, and remain open to be shaped by still unknown facts, and the functioning of these texts as narratives that are persuasive because they are complete.

2. Jean-François Lyotard, *The Post-Modern Condition: A Report on Knowledge* (Manchester: Manchester University Press, 1984), p. 18.

3. For examples, see Beer, *Darwin's Plots;* Misia Landau, "Human Evolution as Narrative," *American Scientist* 72 (1984): 262–67; Jonathan Ree, *Philosophical Tales;* Bruno Latour and C. S. Strum, "Human Social Origins: Oh Please, Tell Us Another Story," *Journal of Social and Biological Structures* 9 (1986): 169–87; Woolgar, "Discovery"; Lynch, *Art and Artifact;* Gregory Myers, "Making a Discovery: Narratives of Split Genes," in *Narrative and Cognition*, ed. Christopher Nash (forthcoming).

The title of a 1985 television documentary, "Science—Fiction" by the BBC played on the ambiguity of such comparisons between science and other narratives, connecting the argument that science was a construction with the popular sense that it must then be false.

Though I will show how each article responds to a previous article, the process of interpretation I describe is not really a dialogue. Michael Mulkay in *The Word and the World* has analyzed an exchange of letters between biochemists with rival theories in terms of conversational analysis. He finds that the authors are aware of conventions of turn-taking parallel to those of dialogue, in which there is one speaker at a time, silences are avoided, and each turn is defined by the next in the sequence. But the exchanges of articles I am studying proceed on principles different from those of conversation (as they are different from the principles of the unequal exchanges between referee and author in chapter 3). The authors do not even address each other; instead they address those who might still be persuadable: a potential audience of herpetologists, comparative zoologists, geneticists, neuroendocrinologists, evolutionists, and ethologists.

The narrative on which this controversy is based is simple enough—one lizard climbs on the back of another, grips the pelvic region of the lower lizard in its jaws, and arches its back so that its cloaca is under the cloaca of the lower lizard. As it happens, all the biologists studied agree that they have seen this sequence of events. But they disagree about its interpretation, about the context in which this narrative should be placed, how it figures in other narratives, such as the story of a career, or the story of the research field as a whole. The significance of the biologists' reinterpretations will be clearer if I give some background to the controversy. But since the background is just what is at issue, I shall present two summaries of the issues involved, both of which explain the materials I have and either of which holds together on its own terms. In order to show the possible differences of perspective the more clearly, I have imagined the accounts of two opposed narrators instead of using the words of any of the biologists I am studying.

One Overview of the Controversy

The *Cnemidophorus* is a genus of lizard that is of special interest because it includes some parthenogenetic species, animals that reproduce from the eggs of the female without any males. Thus they provide an opportunity to study aspects of the control and evolution of sexuality that cannot be separated and analyzed in sexual species. Through the 1970s, researchers carried on increasingly large and sophisticated studies of a number of aspects of the physiology—that is, the bodily processes—of these species. But the researchers were not interested in the ethology—that is, the behavior—of these species. So, in their observations, they did not see anything odd about this behavior. David Crews, a young researcher then at Harvard, had been working on the reproductive processes of another genus of lizard, and became interested in the *Cnemidophorus*. So he brought a comparative approach to the field, and in particular an interest in how behavior is related to hormonal controls. He saw immediately what other researchers had ignored,[4] that these nonsexual lizards, who did not need to mate, sometimes mounted each other, behaving just like the sexual species of the same genus when they mated. And he set out to explain the significance of this paradox. But when he and his co-worker Kevin Fitzgerald tried to publish these findings, they ran into personal opposition from established figures in the field who had, after all, been scooped. Their critics were in a position to block publication at the first journal to which the report was submitted. But it was finally published in the *Proceedings of the National Academy of Sciences*, a prestigious outlet. The article got an unusual amount of publicity, because of its implications for the nature of sexual behavior. Crews and his team of graduate students and postdoctoral fellows then continued with larger-scale studies that, in essentials, confirmed his earlier findings and opened up new perspectives for the comparative approach to the evolution of sexual behavior. What it all shows is how researchers get an investment in one approach to a research area, and then can't see another approach even when there is clear evidence for it. All one can do is keep adding to one's evidence and refining one's methods until all but the diehards accept the new paradigm.

Another Overview of the Controversy

Cnemidophorus are unusually interesting lizards, because they are among the few vertebrates that reproduce parthenogenetically. But they were hard to study at first. The key to *Cnemidophorus* research was the long and difficult process of learning how to maintain them in

4. These two accounts are intentionally slanted. Orlando Cuellar commented on the phrase saying that other researchers had ignored the behavior, suggesting that one could alternatively say "Crews recorded what other researchers had elected to ignore." He also pointed out some particularly slanted phrases in an earlier version, noting that, for instance, the first account, in describing Crews as "a bright young researcher from Harvard," implied "the other guys are old and dumb." But even where the account was not intentionally slanted, there were phrases that either Cuellar or Crews took exception to, showing how loaded an account of a controversy is likely to be. My vagueness about who criticized Crews could create the impression that all the critics I later mention opposed publication of the article; Cuellar points out that he did not review the paper and in fact did not see the paper until after it was published.

captivity. Over the course of about ten years, several research groups contributed findings to this procedure. Those workers who spent years going to the field and then trying to duplicate it in the lab gained a subtle understanding of these lizards—they could, for instance, tell from a photograph whether a lizard was diseased or abnormal in some way, and in a larger sense, they were always aware that a terrarium in Manhattan or Salt Lake City is not the Arizona desert. This work then began to pay off in a number of areas of biology, including genetics, evolution, and physiology, as researchers revealed the mechanisms of parthenogenesis.

At this rather late stage, a young researcher from Harvard, who had worked with a sexual species of lizard (one that is so easy to keep that it is a common pet), and who had just started work on the *Cnemidophorus,* asked an experienced researcher for some additional animals and some advice on maintaining them. The established researcher collected some animals for him and, more important, let the newcomer visit his lab and tap his expertise. But almost as soon as the newcomer got the lizards, and long before he could have gotten publishable results, he seized on a peculiar bit of behavior, noticed in a very few animals, and blew it up into a sensational claim. He saw some lizards mounting others and concluded from this that even unisexual lizards need to mate. Established researchers explained that they had seen such behavior too, but that they could recognize it as unnatural, an artifact of captivity, and so disregard it. The newcomer was encouraged to continue his work, but to wait until he had something more substantial to report. If everything the lizards did was blown up this way, there would be hundreds of articles published, but no progress on the important lines of research. The newcomer pushed his theory even after it was rejected by two reputable journals, and he finally got it published through the influence of a famous biologist who is best known for his work on insects. That would have been the end of it, but because it was about sex, and this newcomer has a genius for publicity, the article was picked up by *Time* as a sort of joke. So some of the established researchers went to the trouble of explaining in print why his theory was ridiculous, drawing on their own extensive records of the lizard's behavior. The newcomer still comes out with articles saying the same thing, but other *Cnemidophorus* workers have better things to do than to refute him again. What it shows is the effect on young researchers of the pressure to publish. This new *Cnemidophorus* researcher has certainly advanced his career. But in the end, this kind of sensationalism and haste doesn't advance science.

These contrasting accounts show what I mean by placing a narrative in context. First, narratives have meaning within other narratives, so that one's interpretation of the lizards' behavior may be linked to one's interpretation of a scientist's behavior, and that interpretation may be linked to one's interpretation of changes in the discipline. On the broadest level the story of a research *field*—in this case, the work by all scientists working on one genus of lizard—is made up of the planned efforts of separate laboratories, each of which can be seen as a separate *project*. The project is made up of individual *studies*, each of which is seen as a sequence of actions leading to a single claim that might be the basis for a published article. And these actions by the researchers are all focused on defining a sequence of actions by the *animals* themselves.

Each of the two overviews I have given moves from the field to the project to the study to the animals, and then back to the study and the project and the field to show the significance, or lack of it, of this narrative of the animal. In the first account, the observation of this behavior by the lizard leads to a study which is part of a larger project on the evolution of sexuality which, if it were successful, would reorient the whole field. In the second account, the lizards are performing the same sequence of actions, but these actions tell about the competence of the researcher, not about nature. The narrative of the study—the discovery or nondiscovery of a behavior—does not follow from that of the lizards, but is explained in terms of the project, which is based on the career goals of the researcher. In either story, the persuasiveness of the interpretation of the animals' actions depends on its place in a larger narrative.

The controversy also makes visible a context that includes all the levels of narrative: the world of texts. This context is usually not acknowledged explicitly: in scientific texts it is assumed that other scientific texts are transparent carriers of a meaning that can be concentrated in the abbreviations of citations. Though scientific texts always cite earlier work, they rarely quote the exact words of another text. As we will see, the participants in this controversy often quote phrases, both reinterpreting the words of their opponents and defining the meaning of their own. Thus the controversy is an arena in which we can see the biologists' own textual criticism.

A participant's reinterpretation of the narrative on one level can lead to an entirely different story on the other levels. There are, of course, many possible interpretations of these narratives, but in this controversy the interpretations come down to just two. When one researcher puts the narrative of another in a new context, its significance is not just altered, it is reversed. This is what I am calling ironic reinterpretation—the perception of a "true" narrative underlying the narrative given. For instance, if one can put a narrative of "normal" behavior in a context in which it is seen as "abnormal," the observation of the behavior is no longer a basis for meaningful statements about the species, but can only be used to raise questions about the correct procedures for maintaining colonies of laboratory animals. Similar reversals are possible with narratives representing the poles of experience or naïveté, open-mindedness or prejudice, discovery or artifact, data or hypothesis.

I shall analyze five published articles.[5] Crews and Fitzgerald published their first article on *Cnemidophorus* in 1980 in *PNAS*. Crews's two major critics, the best-known researchers working with this genus, took the unusual step of rebutting his arguments in print, and Crews and his colleagues responded in print to these criticisms. In 1981 Orlando Cuellar of the University of Utah, who in the early 1970s had shown the chromosomal mechanisms of parthenogenesis, criticized Crews and Fitzgerald in a postscript to a report of a long-term study of the reproductive rhythms of *Cnemidophorus*. Then two years later C. J. Cole of the American Museum of Natural History, who pioneered the physiological study of the genus, and his colleague Carol Townsend published a study of the mounting behavior intended to refute any claim for its reproductive significance, reporting cases and data from their extensive records. The conclusion to their article makes unusually explicit criticism of Crews's group's interpretations of *Cnemidophorus* behavior.

The tone of the postscripts of Cuellar and of Cole and Townsend suggest that the argument cannot be resolved simply by one side producing more data and convincing the other side, though both sides act as if it can. The controversy is over interpretive issues, not over the data. The Crews group did not think it enough just to publish more articles giving more evidence; they also responded to the criticisms directly on two occasions. The response to Cuellar is "Psychobiology of Parthenogenesis," by Crews, Jill Gustafson, and Richard Tokarz (the list of authors gets longer here, so I'll abbreviate this

5. The articles are cited in section 3 of the Reference list. An earlier controversy among *Cnemidophorus* workers can be seen in a review of the ecology and evolution by Cuellar in *Science* in 1977, and criticisms (or "Technical Comments," as *Science* classifies them) published in *Science*, with Cuellar's response, in 1978. This exchange confirms many of the textual features I have described in the controversy over pseudosexual behavior; one can trace in them a hierarchy of narratives, and see ironic reversals, and use of quotations, and the focus on the texts.

as CGT), published in an edited volume, *Lizard Ecology*. The main purpose of the CGT chapter is to present the behavioral inventory that was one of the stated goals of their research when they first began collecting *Cnemidophorus*. This would seem to be a purely descriptive and uncontentious project, but in this controversial context, the inventory that includes a category for "Sexual behaviors"is a challenge to his critics who question the relevance of the behavior Crews describes. CGT acknowledge the their work is subject to controversy only in their concluding section, "Is Pseudocopulatory Behavior in an All-Female Species 'Normal'?"

The response to Cole and Townsend by Michael Moore (then a postdoctoral fellow in Crews's laboratory, and now an assistant professor at Arizona State University) and Joan Whittier, Allan Billy, and Crews (MWBC), combines a report of new findings with a rebuttal of critics' arguments. It was published in 1985 in the same journal that published Cole and Townsend's article (not a usual journal for articles by Crews's group), and is recognizably part of a controversy, not only in its explicit references to Cole, and in its ironic reinterpretation of critics' articles, but in the intensity of its attention to methods and to theoretical assumptions. The sequence of articles in the file is:

1) Crews and Fitzgerald (1980)
2) Cuellar (1981)
3) Cole and Townsend (1983)
4) Crews, Gustafson, and Tokarz [CGT] (1983)
5) Moore, Whittier, Billy, and Crews [MWBC] (1985)

I shall try to show how the narrative of the lizard is constructed in the apparently nonnarrative account of Crews and Fitzgerald. Then I shall trace some of the interpretations and reinterpretations in later articles by focusing on five areas of disagreement, each linked to one part of the standard research article: the claim, the introductory review, the methods, the results and discussion, and the closing. These disagreements involve interpreting what the article said, placing the work in context in the research field, determining whether the study was competent, selecting and interpreting the evidence, and settling the controversy:

1. Each article gives some version of what the Crews and Fitzgerald article was saying, and there is disagreement about the selection and interpretation of the language of the claim.

2. The articles differ, usually in their introductions, in their accounts of the history of the field, and over the place of the Crews and Fitzgerald article in it.

3. The methods sections of the articles expand as previously insignificant details become significant.

4. The interpretation of results often turns on negative results: the lack of evidence that is assumed to indicate the narrative. So the two sides do not meet head on, but through arguments about auxiliary hypotheses.

5. Each article ends by trying to close the controversy—either by declaring the behavior meaningless, or by marking it as an accepted discovery and calling for further study.

Constructing a Narrative of Cnemidophorus *Behavior*

Before we can follow these reinterpretations of the narrative of the lizards, we need to analyze the construction of this underlying narrative itself. We cannot assume that Crews and Fitzgerald simply record a narrative existing in nature—that they are like traditional storytellers, retelling a tale told to them or evident to all observers. If this were the case, other researchers could have seen it immediately, and would have accepted it when it was described to them. The narrative is constructed in the writing of the text. So we need to see how the apparently static form of the scientific article can be used to say, "Once upon a time there were two lizards. . . . " It may seem that narratives like this one would occur only in biology, perhaps only in ethology, the study of animal behavior, and not in other sciences. But it is likely that other sciences deal with other actors, sequences and contexts that are less easily seen by the analyst because they are less easily anthropomorphized.[6]

The narrative on which Crews and Fitzgerald's article and the ensuing controversy is based is a sequence of actions shown in the caption and the photographs in the article's figure 1 (my Appendix figure A3.1). The four photographs combine four positions of lizards into a narrative. To see the photographs as a sequence, one must follow some unstated conventions of interpretation: reading the pictures in the order one would read words on a page, taking the two lizards shown in each of the photographs as the same two lizards, ignoring the third lizard in C, and ignoring the apparent difference in back-

6. For instance, Françoise Bastide of the École des Mines, Paris, presents a clever Greimasian analysis of a *Nature* article in which clay bowls of pipes are actants, in her unpublished paper on "The Semiotic Analysis of Discourse," and in "Une Nuit avec Saturne" she offers an analysis of science reporting in which a satellite is an actant. Bruno Latour treats chemistry texts as trials of strength in *Science in Action* pp. 88–89.

ground as the lizards move around the cage.[7] The caption is essential to our seeing this sequence as a meaningful narrative. First it describes the behavior we are to look for as "sexual" (in quotes). It adds action we cannot see here, such as "lunging attack." It focuses our attention on one aspect of each picture (such as the jaws and foreleg in A), and makes the position in the picture into an action (gripping). And it translates actions into the strictly limited vocabulary of terms used to describe behavior, such as "mounting and riding behavior" or "copulatory posture." This device of attributing terms will be important later in the controversy. Ethologists define a closed set of behaviors for each species; individual lizards may do all sorts of things but a narrative of animal actions, to have ethological meaning, must fit in this repertoire. It is important to recognize it takes work to make this behavior evident, because part of the controversy is over the conditions under which the behavior will or will not be seen.

As it turns out, no one will deny that the lizards can be observed in the positions shown in the photographs in figure A3.1; what they deny is that these actions fit together into a narrative of mating behavior, a meaningfully related sequence of actions. Crews and Fitzgerald must interpret the sequence if it is to be anything more than the random movements of caged animals. The abstract of the article tries to make the narrative meaningful by putting it in a larger context.

> ABSTRACT All-female, parthenogenetic species afford a unique test of hypotheses regarding the nature and evolution of sexuality.

7. Crews comments on the passage in which I analyze the photographs, "These are the same lizards, but as they move about in the cage, the background changes. You imply that I have manufactured the sequence." And he comments, where I say they assembled many different narratives of lizards, that it "implies we never saw the entire thing." I do not mean to say or imply that they fabricated evidence—that these were not the same lizards, or that they only saw part of the behavior at any one time. I use this way of analyzing their evidence to argue that all scientific evidence, even the most apparently straightforward, such as photographs of behavior, requires some interpretive work to make it into a narrative.

Some people who have heard my papers have pointed out that my vocabulary—*construction, narrative, negotiation*—might have connotations of fraud, and Crews's comment shows that scientific readers respond to these connotations. But this danger arises because traditional views of science contrast scientific objectivity, in which the researcher is totally transparent and passive, with fraud or incompetence, in which the researcher is active. I want to show that the production of any scientific knowledge involves social processes that do not fit in the traditional view of scientific knowledge; the traditional view provides no vocabulary for such processes. This only sounds like fabrication or fraud if one assumes that there is some scientific knowledge somewhere that does not involve social construction.

> Basic data on the behavior of parthenogens are lacking, however. We have discovered, from observations of captive *Cnemidophorus uniparens, C. velox,* and *C. tesselatus,* behavior patterns remarkably similar to the courtship and copulatory behavior of closely related sexual species. Briefly, in separately housed pairs, one lizard was repeatedly seen to mount and ride its cagemate and appose the cloacal regions. Dissection and palpation revealed that, in each instance, the courted animal was reproductively active, having ovaries containing large, preovulatory follicles, while the courting animal was either reproductively inactive or postovulatory, having ovaries containing only small, undeveloped follicles. These observations are significant for the questions they raise. For example, is this behavior a nonfunctional vestige of the species' ancestry, or is this behavior necessary for successful reproduction in the species (e.g., by priming reproductive neuro-endocrine mechanisms as has been demonstrated in sexual species)?

This abstract illustrates all the levels of narrative I have outlined: the narrative of the field at the beginning, of the study in the middle, and of the project at the end. Crews and Fitzgerald outline the basic narrative of the lizards in the sentence beginning, "Briefly, in separately housed pairs . . .". The key word in this sentence is *repeatedly;* to construct a narrative, Crews and Fitzgerald had to see many activities over the course of two years as one repeated behavior. A close look at the file shows that the individual animals they chose to exemplify this behavior in any text changed through the course of the study, as they got better examples. The article just notes the first observations of the behavior without details: "In late November 1978, intense social activity was noted in the cages, and daily observations were initiated." The correspondence suggests that this activity was the basis for the report in the first manuscript, sent out in March 1979. Although the published article makes the same claim as the earlier manuscript, the claim is now based, not on whatever was observed in November, but on a much larger number of animals collected in June 1979 and observed in July and August 1979. Observations of two other species are also described, without any dates given in the article. Their numbers are small, so the observations might not have been publishable in themselves, but after the *C. uniparens* narrative, they can serve as supporting data.

Crews and Fitzgerald make their interpretation of the basic narrative of the lizards' behavior by juxtaposing it in their study with two other narratives: the normal mating behavior of a pair of sexual lizards ("behavior patterns remarkably similar to . . . ") and the unisexual lizards' own reproductive cycles ("in each instance, the courted animal was reproductively active . . . "). In the article, the parallel with sexual species is made by the photographs in figure A3.1, showing unisexual lizards in what the captions interpret as the same positions as those of sexual lizards. (A later article puts three photographs of a pair of sexual lizards mating alongside three photographs of *C. uniparens* to make the point more explicitly.)

The second juxtaposition that turns actions into a meaningful narrative is made by a table that relates the behavior of animals to their reproductive state (my figure A3.2). This may not seem to be a narrative, since all elements of time are removed to make the various observations of lizards simultaneous. But by linking the reproductive state of each lizard to its behavior, Crews and Fitzgerald place the behavior, over the course of a few minutes, in the larger narrative of the reproductive cycle over the course of months. This narrative of the study depends on the narrative of the animal that was earlier created, for the malelike or femalelike behavior must be established before it can be related to reproductive state. The activity of Crews and Fitzgerald and their coworkers is confined to the notes to this table, which tell us, for instance, that they determined the reproductive state through three different procedures; dissection and observation of the size of follicles, or recording of the laying of eggs, or the palpation of the undissected animal. For the table to be coherent, these quite different procedures must be assumed to describe the same condition. The table creates a narrative by selecting details, as well as by organizing them. For instance, it shows body length (to demonstrate that it is not important) and the method of determining the reproductive state for each animal. The table is also significant for what it does not include. A critic of Crews's findings had asked about length of time the animals were in captivity, and about egg-laying records, and had asked for comparisons to lizards who had not engaged in this behavior, with the implication that the inclusion of these data might lead to a quite different story.

Each of these devices, the abstract, the figure, and the table, compiles a series of momentary observations of lizards into a narrative of the study in which the events take on a larger meaning. The narratives of the lizards and the narrative of the study are framed in the text by a narrative on the level of the research project: the creation of a discovery.[8] This narrative, too, requires selection and ordering. The

8. Woolgar, "Discovery"; Myers, "Making a Discovery."

first sentence of the abstract places previous work in the field in the context of Crews and Fitzgerald's own questions about "the nature and evolution of sexuality," rather than in the context of other research projects on the lizards' cytogenetics and ecology. The second sentence of the abstract identifies a significant lacuna in the literature, which Crews can fill: "Basic data on the behavior of parthenogens are lacking". They end the abstract with new research questions raised by his work. The discovery is presented as creating a new research project, one that will pursue the questions raised by the observation of this behavior.[9] Crews treats his work as if it were parallel to earlier studies by other researchers, citing their work prominently (as in the caption to figure A3.1) to show his procedures are unexceptionable. The irony of this reinterpretation is that behavior that other researchers had treated as insignificant—apparent mounting in a unisexual species—is now reinterpreted as significant. As one might expect, other researchers are not happy to have their studies reinterpreted in this way.

Reinterpreting the Narrative of Cnemidophorus *Behavior*

The Claim

The claim of an article, its main contribution to knowledge, is usually taken to be unambiguous, so that an unqualified reference like "Crews and Fitzgerald, Proc. Natl. Acad. Sci. USA 77 (1980) 499–502" can convey it. As I have noted in the introduction, a large body of sociological scholarship is based on the links such references make. But Nigel Gilbert has shown how a number of different sentences in an article could be taken as the claim. John Swales has summarized the research on citation context analysis and shown how these references serve a number of textual functions besides simply providing a structure of knowledge on which to base further work.[10] And I show in chapter 2 that the claim of an article can change in the course of revision and review. What we see in these texts is that the readings of a claim can vary widely among participants in a controversy. It is difficult to see this variance in most citations, because they do not quote the words they take as a claim; the particular words of a scientific article, unlike those of, say a literary critic, are not important in

9. See John Swales, *Aspects of Article Introductions* (Birmingham: Aston University, 1981), for discussion of these moves.

10. Gilbert, "The Transformation of Research Findings into Scientific Knowledge": Swales, "Citation Analysis and Discourse Analysis."

later references. In fact, quotations might nearly always be taken as a sign of trouble; something must be focusing attention on the text itself, which usually vanishes from sight in the accumulation of claims.

Cuellar's citation of Crews and Fitzgerald seems straightforward enough:

> Recently, Crew and Fitzgerald (1980) reported the discovery of copulations among several all-female species of *Cnemidophorus,* including *uniparens,* and proposed that such pseudocopulations may be necessary for successful reproduction in the species. ("Long-term Analysis," p.99)

Latour and Woolgar discuss modal shifts in presentation of claims in *Laboratory Life;* their point is that verb phrases like "reported the discovery of copulations" and "proposed that such psuedocopulations may be necessary" imply less of an attribution of fact than would phrases saying Crews and Fitzgerald "discovered copulations" and "showed they were necessary." To remind us of the agency of the researcher, the author, is to weaken the claim.

Similarly, Cole and Townsend's citation signals that they will question the claim by focusing on the exact words used by Crews and Fitzgerald. Cole and Townsend's use of the exact words is an example of what Dan Sperber and Dierdre Wilson call "echoic speech." Sperber and Wilson give this example: "*He:* 'It's a lovely day for a picnic.' [They go for a picnic and it rains.] *She* (sarcastically):'It's a lovely day for a picnic, indeed.' "[11] Cole and Townsend's use of Crews and Fitzgerald's words has a similar, if less obvious, effect of irony.

> Recently, Crews and Fitzgerald (1980) reported that captive females of unisexual species of lizards exhibit 'behavior patterns remarkably similar to the courtship and copulatory behavior of sexual congenerics'. Although other investigators had observed this also (Schall 1976, Werner 1980; Cuellar 1981; personal observations), only Crews and Fitzgerald (1980) suggested that homosexual behavior is normal in the reproductive biology of unisexual lizards. For whiptail lizards, they stated that: (a) 'In each instance', in *Cnemidophorus uniparens,* the female exhibiting malelike courting and mounting was 'reproductively inactive', (b) the female being courted was

11. Dan Sperber and Deirdre Wilson, *Relevance: Communication and Cognition* (Oxford: Basil Blackwell, 1986), p. 239.

> 'reproductively active', (c) such behavioural interaction was not seen among 'any females of sexual species', and (d) such interaction among unisexual species of *Cnemidiphorus* 'may be required for successful reproduction'.

Cole and Townsend select words of the Crews and Fitzgerald article to define the claims they will refute; they end this passage with the statement "Our own observations . . . contradict these suggestions." The implication of the quoted words is that Crews and Fitzgerald have stated their case much too strongly ("in each instance" "any females of sexual species") or that their terminology is vague ("reproductively active") or that they have gone out on a limb of hypothesis ("may be required for successful reproduction") and now they are going to get their comeuppance. Their claim is paraphrased as well as quoted; only Crews and Fitzgerald have "suggested that homosexual behavior is normal." The use of the word *homosexual,* which does not occur in Crews' articles, implies a sensationalism on his part.

As in some controversies in literary criticism, both sides treat the original text as unambiguous and the interpretations put on it by the other side as selective, overingenious, and transparently motivated. CGT respond to Cuellar's interpretation of Crews and Fitzgerald's language with some reinterpretations of their own.

> Cuellar (1981) has stated that Crews and Fitzgerald (1980) "proposed that such pseudocopulations may be necessary for successful reproduction," and others have echoed this statement. But this is a misinterpretation of that paper.

The disagreement shows that the reduction of a four-page article to a one sentence statement is not all trivial or automatic. CGT themselves reinterpret the context, the significance, and the strength of the claim in the earlier paper. First, they define it as an "initial report".

> The purpose of that initial report was to document the alternation of male-like and female-like sexual behaviors during specific stages of the follicular cycle in 3 unisexual *Cnemidophorus* species.

In calling it "initial," CGT imply that it is to be read in conjunction with the series of papers that followed it and refined it. CGT restrict the significance of Crews and Fitzgerald's original paper to its documenting that malelike and femalelike sexual behavior alternate. Then they offer a version of the claim:

> It [Crews and Fitzgerald] concluded that "It is likely that social interactions play an important role in in the reproductive biology of parthenogenetic *Cnemidophorus*" and raised the question of whether "this behavior may be necessary for successful reproduction in the species (for instance, by priming reproductive neuroendocrine mechanisms) as has been demonstrated for some sexual species." Obviously, Cuellar and others have chosen to interpret this to mean that Crews and Fitzgerald (and the present investigators) believe pseudocopulatory behavior to be essential for reproduction. We would like to emphasize that this is not our intention. Rather, we are suggesting that the presence and behavior of conspecifics may act as a neuroendocrine primer and facilitate reproduction in parthenogenetic lizards as does male courtship in sexual lizards.

The reading of Crews and Fitzgerald by CGT, like the readings by Cole and Townsend or Cuellar, is rather selective; CGT combine the first sentence of the last paragraph with the last sentence of the abstract to make the article claim that social interactions are important, and that they are important only in priming, in facilitating rather than causing reproduction. This claim is much weaker, and much easier to support, than the claim Cuellar atributes to them, that such behavior is necessary.

As Cuellar and Cole show, it is possible to get a number of other claims out of that paper. It is easy to see where Cuellar and others get their reading; the last sentence of Crews and Fitzgerald's article, taken out of context, says almost exactly what Cuellar says they say: "malelike sexual behavior in parthenogenetic *Cnemidophorus* may be required for successful reproduction." This seems to be a fairly definite answer to the question asked by Crews and Fitzgerald earlier in the article: "Is it necessary for sexual reproduction?" How, then, can CGT argue Crews and Fitzgerald don't mean pseudocopulation is essential for reproduction? They go on in another paragraph to define successful reproduction, not as the hatching of eggs, but as the hatching of eggs at a "normal" rate. Also they imply that to understand the claim properly one needs to consider the research context, as CGT do in the last sentence of the paragraph: "It has long been known that eggs laid by isolated unisexual lizards will hatch, (Maslin, 1971), a finding confirmed in our laboratory." It would seem that neither side really thinks the stronger claim—that sexual behavior is required for any reproduction—is being made, for neither side actually tests it with the simple experiment of keeping a lizard isolated and seeing if it lays and hatches eggs. Even Cole and Townsend, arguing this point, have to refer to such an experiment with an entirely different genus—not, presumably, because they couldn't show it with *Cnemidophorus*, but because such an experiment would, in the current context, be trivial. When CGT note that the point was established by Maslin ten years earlier, they imply that Crews and Fitzgerald could not have been ignorant of these established results, and could not have been contradicting them.[12] Crews and Fitzgerald couldn't have meant what Cuellar says they meant, because everyone knows that isolated eggs will hatch, so they must have meant something else. We shall see other examples of assumed contexts for the interpretation of texts in the controversy over *Sociobiology* (chapter 6). The interpretation of a claim depends on the place it is given in the narrative of the research field, especially in the version of this narrative given in the review of research that opens the article.

The Introductory Review

Each article in the file places the research of Crews and Fitzgerald in the larger context of the issues of importance in the research field. Crews and Fitzgerald, as we have seen, claim significance for their observation by showing it fills a gap in the literature, which had emphasized physiology and ecology but overlooked studies in behavior. The introductions of the Cole and Townsend article and the MWBC article show how the different sides of the controversy construct different views of the research field.

Cole and Townsend present a view of the field in which Crews and Fitzgerald are describing an artifact, not a discovery. Artifacts in ethology are somewhat different from those in, say, microscopy (which Lynch describes in *Art and Artifact*), where researchers try to determine which features of an image are the result of experimental manipulation or instrumental procedures (the artifacts) and separate them from the features of the image that are taken as a true representation of nature. The artifact is behavior that falls outside the narrative of normality—in this case, behavior that results from confinement in terraria in a laboratory, and thus does not reflect the way the animals behave in the desert. So Cole and Townsend grant in the first sen-

12. Strikingly similar disagreements about the claim can be found in the "Technical Comments" and response after Cuellar's 1977 *Science* review. For instance, Cuellar responds to critics' versions of his claims, which he sees as oversimplified, by referring, just as Crews does, to a well-known background of research against which the claims must be interpreted.

tence of their abstract that the behavior exists, but state in their second sentence that it is abnormal.

> Abstract. In captivity, females of parthenogenetic species of whiptail lizards (*Cnemidophorus*) occasionally mount other females and behave as if attempting to mate. This occurs under crowded conditions, and probably is not related to reproduction.

Cole and Townsend then go on to outline a history of the field in which Cole's own work is central, reminding the reader of his long experience with this line of research.

> The unisexual species of reptiles that reproduce parthenogenetically may be the only vertebrates in which individual females normally reproduce independently of males (Cole 1975; Hardy & Cole 1981; see Downs 1978 for possible examples in salamanders). Consequently, all aspects of their reproductive biology merit attention. Recently, Crews and Fitzgerald (1980) reported that captive females of unisexual species of lizards exhibit 'behavior patterns remarkably similar to the courtship and copulatory behavior of sexual congenerics'. Although other investigators had observed this also (Schall 1976; Werner 1980; Cuellar 1981; personal observations), only Crews and Fitzgerald (1980) suggested that homosexual behavior is normal in the reproductive biology of unisexual lizards.

Like Crews and Fitgerald's article, this acticle begins by saying that the animals are so important that one must pay attention to all aspects of research on them. But the reference to Crews and Fitzgerald is followed immediately with the qualification that their report is not a discovery, because the behavior has often been observed before, and that their interpretation of the behavior is idiosyncratic, a diversion from the main line of research, in which they have isolated themselves from the consensus of the field.

The title of the MWBC article—"Male-like behavior in an all-female lizard: relation to ovarian cycle"—has two parts forming a narrative of the research project, one taking the existence of the behavior for granted, as a given topic, and the other adding a new contribution. The first sentence defines a line of research conducted entirely by Crews's group: "Recent observations of copulatory-like behavior in all-female species of parthenogenetic lizards have emphasized the dual functions of sexual behavior (Crews and Fitzgerald 1980; Gustafson and Crews 1981; Crews 1982; Crews et al. 1983)." So far, one would not think this a matter of controversy, unless one notes that all the work cited was done in one lab—their own. The MWBC authors provide an ironic reinterpretation of the field—their own work as well as that of their critics—that must be exasperating to anyone trying to disagree with them. In it they simply assume pseudocopulatory behavior as a fact and as a term, and go on to the need for further studies of it.

> Copulatory-like behaviour in unisexual lizards was first described by Crews and Fitzgerald (1980). The occurrence of this behaviour, hereafter called pseudocopulation, has recently been confirmed by other workers (Werner 1980; Cuellar 1981; Cole and Townsend 1983). Crews and Fitzgerald observed that male- and female-like copulatory behaviour was exhibited in separate phases of the ovarian cycle: female-like roles occurred only during vitellogenesis and male-like roles occurred only during pre-vitellogenesis or after ovulation. This led them to hypothesize that individual animals alternate between male-like and female-like behaviour as the ovarian cycle progresses. Recently, this interpretation has been challenged by Cuellar (1981) and Cole and Townsend (1983). However, all reports so far have been descriptive studies, which employed small sample sizes, thereby precluding quantitative analysis.

MWBC incorporate the evidence given by Cuellar and Cole and Townsend into the Crews case, even though these critics gave the evidence only to show that the circumstances refuted Crews's claim. Note that in appropriating these findings, MWBC also rename them in terms of their own terminology, *pseudocopulation,* which Cuellar and Cole certainly would not accept. There is no need for the term if the behavior is merely an artifact of captivity.

MWBC separate the relation of the behavior to the ovarian cycle, which they take as "observed," from the claim that individual animals alternate roles that Crews and Fitzgerald were "led . . . to hypothesize." This hypothesis is the issue that MWBC will address and extend in the paper. On this "interpretation," the same authors who are used for support in the previous paragraph serve as antagonists. But all earlier articles, including those of Crews's group, are made preliminary to the present study, in which "we report . . . the first quantitative analysis of the relationship of copulatory-like behavior to ovarian states in a unisexual lizard." Thus they seem to make the earlier controversy over Crews and Fitzgerald irrelevant, based as it was on the insufficient data then available.[13]

Methods

One striking characteristic that sets these texts apart from noncontroversial articles is the expansion of the Methods sections during the controversy. The Crews and Fitzgerald article can get its methodological details in tiny print in the caption to its figure 1, citing Cole for further details. But the MWBC article, five years later, gives a remarkable amount of detail about the regimen of care, observational procedures, and categories of reproductive state. Latour and Bastide ("Writing Science") show how methods sections, usually thought by students to be dull formalities, become crucial in controversies. One would expect that, under the stress of controversy, studies might become more elaborate, and might be done on a larger scale to be more persuasive. But what also seems to be happening here is that the list of relevant information grows each time one side questions the technique of the other. As in the gravity waves work that Collins describes in *Changing Order,* in which some researchers were trying to refute an apparently bizarre claim, there is no objective standard for what would constitute replication or refutation of the original observations. A modification that may be seen by a researcher as a minor variation or an improvement in the apparatus may be seen by another researcher as invalidating the evidence of that apparatus.

When Crews and Fitzgerald say in their first article that behavioral data are lacking, they raise the methodological issue of what one has to do to see behavior, and thus focus attention on the skills of the established researchers in the field they themselves have just entered. Cuellar defends himself from the implication that he missed this behavior by referring to his long experience.

> During the last decade I have monitored the development and laying of nearly 1000 clutches from captive *C. uniparens.* Since my studies have required precise knowledge of ovulation and oviposition times, copulatory behavior would have revealed itself as a most conspicuous feature of the reproductive cycle of the species.

13. The 1977 controversy in *Science* also included an exchange between Cole and Cueller over which articles should form the basis for future work, both of them referring to their own review articles. See also the comments in chapter 6 on the construction and reconstruction of a research tradition in the sociobiology controversy.

The language is unusual for a scientific article because, as an examination of the subjects of his sentences suggests, it focuses attention on him and his studies, rather than on the animal. "Decade" is a word used in the article only to refer to the passing of human time, of careers, not to the cycle of lizards or the measured time of experiments. And Cuellar points out his "precise knowledge" of the reproductive cycle in a way he would hardly do unless he felt his observations had been challenged. His choice of verb defines a model of observation in which the object makes an imprint on the passive watcher; the behavior "would have revealed itself."

The Cole and Townsend article is in the form of a refutation of Crews and Fitzgerald. But instead of planning an experiment along lines suggested in the article they want to refute, Cole and Townsend reinterpret the data they had gathered for other purposes, and present this reinterpretation as the equivalent of Crews and Fitzgerald's study, or rather as an improvement on it. Indeed, it could be argued that there would be no point in a replication, since both sides agree that the behavior occurs, and they disagree only about its significance (Collins, *Changing Order*). In order to show that their procedures would not miss the behavior, Cole and Townsend must go into surprising detail, and like Cuellar, they refer to their own skills more directly than they might in a noncontroversial article.

> Lizards in our laboratory colonies of unisexual species have been reproducing since 1972 (Cole and Townsend 1977; Townsend 1979). Each animal is uniquely marked for individual recognition and notes are kept regarding genealogy, dates of hatching and death, oviposition, cagemates, and other observations. Although our procedures were designed to investigate non-behavioural aspects of reproduction, genetics, and systematics, we also recorded male-like behaviour among these animals whenever it was seen. Lizards judged to be gravid had a characteristically swollen abdomen and usually oviposited within a week of the observations recorded. Since most of these lizards were kept in one of our offices, they were under close, although casual, observation. We provided nearly all their care ourselves, and because they are diurnal, it is not likely that we missed much behaviour pertinent to this report.

Like Cuellar, Cole and Townsend stress the duration and detail of their observations, in contrast to the short term of Crews and Fitzgerald's work. The long list of categories in which notes were taken (such as genealogy) may not be entirely relevant to the research reported; it does not prepare the reader for their argument but stands as testimony to the detail of their study and to the selectivity of the records kept by Crews and Fitzgerald. It is unlikely that outside the context of public controversy they would need to mention that the lizards were kept in their offices, or that they cared for the animals personally. Twice they refer to the fact that they were not specifically looking for this behavior, but three times they comment on the closeness of their watch. That they can use their previously recorded observations in this new narrative at all shows that exactly the same events can be used to make two different studies in two different research projects.

CGT comment on methodology in a response to Cuellar that offers a reinterpretation of his research project in which his diligence and experience count against him. They take his observation of the behavior as confirmation, and make his failure to interpret the behavior as mating into an indication of his preconceptions. Again the quotation of a phrase signals an unusual use of the text of another researcher. In this case, CGT quote Cuellar because they are appropriating his findings as confirmation of theirs:

> In support of this interpretation, Cuellar states that he has "observed such behavior in *C. uniparens* and [unisexual] species in the laboratory for fifteen years, but only sporadically." But it is significant that Cuellar, as well as other workers, has observed male-like behavior in parthenogenetic *Cnemidophorus*. That these observations have gone unreported in previous studies should not be too surprising. Since the function of these courtship and copulatory behaviors is not obvious, these workers most likely felt that this behavior was an abnormal manifestation of captivity. Preconceptions, however, guide perception, and one does not very often see what one is not looking for.

The first sentence repeats Cuellar's ironic turn of Crews and Fitzgerald's discovery of the behavior—the behavior happens, but they have not discovered it and it is not normal anyway. Then they do an ironic turn on the ironic turn, holding that the important assertion in Cuellar's article is that he *did* see the behavior—so he confirms Crews and Fitzgerald in spite of himself. They propose that there is a need to explain why this behavior went unreported, why the observation was not, until Crews and Fitzgerald, defined as a discovery. And they explain, in the terms I am using, that the narrative of the project comes first, that one has to have an explanation in mind before one will see the narrative of the animal the way CGT do. They propose, not a methodology, but a philosophical rule for the whole field, a reinterpretation of the way researchers do their research: "preconceptions guide perception." This series of reinterpretations follows from the need to explain why experienced researchers would construct two different narratives for what these lizards are doing.[14] Cuellar and CGT are using two different conceptions of what makes a good observer, Cuellar saying observation depends on experience and CGT saying it depends on theoretical orientation. Both imply criticisms of personal scientific practice that are very rare in the form in scientific articles.

The methods section of the MWBC article shows that it is intended to refute Cole and Townsend, and not just supersede their data; it is much more detailed than earlier articles, and responds to the criticisms Cuellar and Cole and Townsend had implied. One peculiarity that suggests this methods section is a response to previous criticisms is the recurrence of the word *careful* when they say that "careful notes were taken" of every pseudocopulation among the experimental animals, "we also kept careful notes" on pseudocopulation among other animals in the lab, and "careful records were kept of egg-laying dates." The word carries no information, for one cannot imagine MWBC reporting that they kept *careless* notes, but it does make sense as a response to Cole and Townsend's claim for the detail of their records and observation.

But the methods section does not give so many details just to demonstrate their care; MWBC focus on details that have a rhetorical purpose in their response to Cole and Townsend. For instance, MWBC give a great deal of detail on their terraria, since Cole and Townsend and Cuellar had said Crews and Fitzgerald were observing an artifact of confinement. The procedures for care of the lizards from Cole's and Cuellar's earlier articles are cited. Exact dates and duration of observation are given (this is a point on which CGT had criticized Cuellar). Observation of behavior is for the first time related to ethograms, more formal and rigorous repertoires of behavior, suggesting that the issue of categorization of behavior ("basking" or "arm waving"), raised by the postscript to Cole and Townsend, has been resolved. MWBC also describe in detail their methods for determining and classifying reproductive state, which they argue will show a correlation where Cole and Townsend's gravid/nongravid distinction did not.

14. Gilbert and Mulkay discuss such explanations in *Opening Pandora's Box,* pp. 63–89.

The response to Cole and Townsend in MWBC's discussion makes it clear why their own methods section has become crucial and has become long. First they attack Cole and Townsend's methods of observing and classifying the state of the lizards:

> Cole & Townsend may have reached this conclusion because their method of assessing reproductive state by visual examination of abdominal distension is not adequate for making the critical (see Fig. 1) distinction between animals that have large yolking follicles and those with oviducal eggs. This distinction can be made only by palpating the abdomen as described by Cuellar (1971).

The methods for observing reproductive state that Cole and Townsend used for earlier studies are held to be insufficient for this new area of research; the preferred methods are supported by a reference to another of their critics.

MWBC's criticism of Cole and Townsend's omissions shows the rhetorical intent of the comments in their own methods section on cage size. In an ironic turn MWBC say Cole and Townsend neglect the data necessary to confirm their own hypothesis, whereas MWBC have given the relevant data in their methods.

> This conclusion [that the behavior has no effect on reproduction] is based solely on egg-laying records; no information on the reproductive history or social environment of their captive animals being provided. Cole and Townsend argue further that pseudocopulation is an artifact of crowded conditions in captivity, yet they present no data to support this hypothesis. They do not give the dimensions of the cages used, nor the number of animals housed per cage. In fact, by their own admission, their experiments were designed to investigate 'nonbehavioural aspects of reproduction, genetics, and systematics.'

When every detail of method is questioned as closely as this, both sides must present cases for each procedure they use. Previously ignored aspects of the study—what room the cages were in, who fed the animals, the size of the cages, which animals were in each cage—now become potentially significant.

Negative Results

One striking feature of the interpretations of the *Cnemidophorus* narrative is the weight given to negative results, not to evidence of the narrative, but to missing evidence that would be needed to support the narrative. This might be seen to support a Popperian interpretation, in which the rival researchers seek to falsify the hypothesis of Crews and Fitzgerald. But the ingenuity with which auxiliary hypotheses proliferate suggests that we are seeing a pattern of ironic rhetoric, a tendency to respond not by refuting, but by reversing the rival claim. The arguments do not generate new data on the same issues, but generate further issues.

Perhaps I can clarify what I mean by negative evidence by referring to the case study of a better-known researcher in a better-known controversy: Sherlock Holmes in A. Conan Doyle's *The Sign of Four*. In chapter 6, Mr. Athelney Jones, the hapless police inspector, is spinning out a narrative of the death of Bartholomew Sholto. He is, like all the police in these stories, totally off the track, and Holmes must set him right.

> "Ha! I have a theory. . . . What do you think of this, Holmes? Sholto was, on his own confession, with his brother last night. The brother died in a fit, on which Sholto walked off with the treasure! How's that?"
>
> "On which the dead man very considerately got up and locked the door on the inside."
>
> "Hum! There's a flaw there . . . " (p. 189)

The narrative proposed by Jones requires a door locked from the outside; that it is locked from the inside indicates the theory is wrong, and that there must be an alternative theory, which Holmes will eventually reveal to us. In the same way (without implying that Crews is as dim as the police in the Sherlock Holmes stories), Crews's critics suggest that there is evidence that would have to be there to support his case, but is not.[15]

The most powerful piece of this negative evidence is simply that no one, including Crews and Fitzgerald, has seen this behavior in the field; Cole and Townsend mention that the most thorough study of their behavior in the wild does not include it. The response by Crews's group is not an offer of evidence of copulation in the field, but an ironic

15. All the studies of controversies in note 1 to this chapter refer to the importance (or disregard) of negative evidence, so the focus on negative evidence here would seem to be a general feature of scientific debate, and not, as it might seem, a sign of the trivialization of debate.

reversal questioning the appropriateness of the evidence called for. They question the critics' contrast of the field (as the place of pure behavior) versus the laboratory (as a place of confined and abnormal behavior). CGT argue that the difficulty of observing *Cnemidophorus* invalidates the negative evidence about such matings in the field.

> That male-like sexual behavior has not been observed in unisexual *Cnemidophorus* lizards in nature does not mean that it does not occur. Anyone who has worked with cnemidophorine lizards in the field knows how difficult they are to observe. *Cnemidophorus uniparens* are extremely active foragers and spend much of their time above ground in thick mesquite and creosote bushes. They are wary of humans and, if approached too closely, will retreat quickly into extensive burrow systems. Furthermore, the literature indicates that matings even in sexual *Cnemidophorus* are observed in nature only infrequently.

The ironic turn of CGT is to move from the question of the observation of this behavior in parthenogenetic species to the problems of observation itself, as shown by the difficulty of seeing the sexual species engaged in mating behavior. If normal mating behavior is *not* observed in sexual species in the field, then the fact that it is *not* observed in parthenogenetic species is insignificant.[16] Instead of responding to negative evidence with positive, they respond with other negative evidence that works against the first negative evidence. The argument is like that which Darwin makes to explain the lack of continuous fossil record of gradual evolution, when he describes the record as a mutilated book (*The Origin of Species* p. 316). But to make this argument from what they *don't* see they must present a case for themselves as field observers, and they do this with the same sort of detail Cuellar and Cole and Townsend give in their defenses of their observations. The descriptions of the bushes in which the lizards are found add authenticity. They use vague adverbs that are rare in this kind of article (*extremely* active, *very* wary, retreat *quickly*). The effect is to create a visual image of the real desert where the lizards are found, and to suggest that the authors are old hands when it comes to field experience.

16. In reviews of the later paper by Crews and Moore, several *Cnemidophorus* workers insisted that it was possible to observe sexual *Cnemidophorus* lizards mating, even though no such observations had been published (of course, such observations would hardly be surprising enough to merit publication). But none of these reviewers uses this denial of the negative evidence as a basis for rejection of Crews and Moore's major claim.

Another piece of negative evidence concerns the stage of the behavior in which the mounting lizard grasps the mounted lizard in its jaws. Cuellar argues the lizards should show bite marks if they have been mounted.

> I have observed such behavior in *C. uniparens* and other species for 15 years, but only sporadically. Moreover, only in rare instances have we observed copulation bites among nearly 2000 individuals collected in the wild. The abnormal constraints imposed by captivity result in a variety of bizarre behaviors of which female-female matings is but one.

The argument here is that the mounted lizard should have a bite mark after copulation, that there are few bite marks in the field, so they could not have been mounted in the field, so the behavior must occur only in captivity.

CGT respond to Cuellar's negative evidence, not by providing evidence of bite marks on parthenogenetic lizards, but by ironically reversing the argument, and pointing out that similar negative evidence would apply to sexual species. This does not prove that sexual species don't mate, but suggests that the sign is not a natural inscription of mating.

> He [Cuellar] provides no information about the frequency of such marks in sexual versus unisexual *Cnemidophorus.* Examination of 1,000 female adult *C. tigris,* a sexual species, collected during the breeding season and deposited in the Museum of Vertebrate Zoology, University of California, Berkeley, revealed that only 3 percent had marks on the back and side; further, the same frequency of males (N=1,100) possessed such marks (Crews, unpublished data). On the basis of our behavioral observations of both sexual and unisexual cnemidophorine lizards, we would suggest that these marks reflect interspecific aggression, predation attempts, or accidents.

CGT have been pushed into a rather strange piece of counting by Cuellar's criticisms. They seem to recognize that part of the persuasiveness of Cuellar's point is simply the large numbers he can muster. If they did not need to make this argument, it is hard to see why they would look at more than a thousand sexual lizards as part of a study of a unisexual species. The response to Cuellar's reinterpretation of their animal narrative is another parallel of sexual and parthenogenetic lizard behaviors. Just as Cole and Townsend had reinterpreted

Crews by granting the behavior, but relating it to aggression in laboratory conditions, CGT find the marks that Cuellar observes, but relate them to the rough life in the field. CGT present this ironic turn, not as a reinterpretation, but as a request for further information ("He provides no information . . . ").

The statistics presented in tables in Cole and Townsend's article present another sort of negative evidence, apparently denying a reproductive role for the behavior. Cole and Townsend conclude, "Our observations suggest there is no correlation between the reproductive states of the mounting lizard and the mountee (Table 1)." This seems a clear refutation, but the categories presented in their table and in the table in Crews and Fitzgerald's article are not quite comparable. Crews and Fitzgerald used the terms previtellogenic, preovulatory, and postovulatory, whereas Cole and Townsend categorize their lizards as gravid and nongravid; the difference, as we have seen, becomes a matter of controversy in the methods sections. Cole and Townsend demonstrate that the behavior "is of no obvious benefit to their reproduction" by counting the eggs laid in each clutch. A reader in Crews's lab notes in the margin at this point, "interval between clutches" suggesting a possible criticism of the use of the number of eggs as a measure of reproduction.[17] As in the gravity waves case that Collins studied (*Changing Order*), when the phenomenon is in question, there is disagreement over what counts as a competent experiment. The controversy moves from the phenomenon, to ways of observing the phenomenon, to checks on ways of observing the phenomenon in what Collins calls "the experimenter's regress."

Cole and Townsend provide another kind of negative evidence with observations of the behavior occurring in contexts in which it could have no reproductive function. The following passage, for instance, is a reinterpretation in which the behavior is observed but is rendered meaningless because it does not correlate with reproductive state as Crews and Fitzgerald said it would.

> In this regard, the two following sets of observations are interesting because they are the occasions on which a female was observed to mount all other inhabitants of her cage in one day. One of these

17. This sentence in an earlier draft read, "A reader in Crews' lab notes in the margin at this point, 'interval between clutches,' suggesting the possibility that these lizards might lay eggs more frequently, if not in larger numbers.' " Crews's comment in the margin of my paper here emphasizes the importance of the methodological differences between his group and Cole's. "The laying of eggs has nothing to do with it," he says, "it is the size of the ovary that is all important."

> mounters was a non-gravid *C. neomexicanus*, which mounted three *C. uniparens* (two gravid, one non-gravid). The other mounter was a gravid *C. uniparens*, which mounted two *C. exsanguis* (both non-gravid) and one *C. uniparens* (gravid). No correlations are indicated.

As the last sentence suggests, the intended significance of these accounts is to show that the behavior has nothing to do with the laying of eggs—it could not be related, if both gravid and nongravid lizards mount, and if the mounted lizards are both gravid and nongravid. Instead, it supports the counter-argument that the behavior is an artifact of captivity. It is significant that they use a narrative form of evidence, instead of sticking to a quantitative argument; both Cole's group and Crews's group seem to realize that cases can be more persuasive in some contexts than large numbers. MWBC examine in detail the alternations of role by just three lizards, and Crews and Moore have made an attractive figure illustrating this case for use in other articles.

The use of negative evidence can be complex, for it requires the reader to imagine a complex series of causes and effects, or rather of noncauses and noneffects. This complexity is apparent in Cole and Townsend's argument that the behavior is a form of territoriality induced by captivity.

> It may be significant in this respect that in 60% of our observations of females mounting females, the partners were not conspecific (Table I), though in most cases a conspecific female also was present in the cage. In addition, in 50% of our observations on captives (Table I), the mounter was *C. uniparens*, although Hulse (1981) reported no such behaviour in a field study of this species, which included observations through two summers (7 months). Hulse (1981) also stated: "*Cnemidophorus uniparens* exhibited no signs of territoriality". We suspect that confinement in captivity enhances this activity. In this regard even Werner's (1980) observations on free-living geckoes are pertinent, as the animals were in a dense population in an artificial environment (human habitations).

The evidence that the lizards often mounted members of other species confirms that they do mount in this way, but suggests that it has nothing to do with reproduction. The argument Cole and Townsend then make by focusing on *C. uniparens* is rather complicated. One would think that the evidence that *C. uniparens* are the most frequent mounters, and that they are not observed to be particularly territorial, would work against the claim that the behavior was territorial and for Crews and Fitzgerald's claim that it was related to reproduction. But Cole and Townsend go on to say that the species might then become territorial in captivity, suggesting again that the behavior is an artifact. It is, in their view, unrelated to the lizards' natural territorial behavior as it is unrelated to their natural reproductive cycles. The one piece of evidence for such mounting in the wild, in another genus, is turned so that it is evidence for such behavior being the result of captivity. The geckos live in people's houses; this characteristic can be reinterpreted so that what Crews calls "wild" (implying naturalness), Cole and Townsend call "human habitations" (implying artificiality).[18]

Closure

The analyst with only these texts for evidence would think that the controversy was always just about to end, for each article ends with a reassuring note of closure, setting out some firm grounds to justify further work or deny any need for it. There is a rhetorical difficulty in such closure, for scientists are suppose to seem open to further questioning, especially at the end of the article, conventionally the place for references to further work. The Crews and Fitzgerald article, for instance, ends with a whole paragraph of questions. Cuellar's ending, if read literally, says that he is open to further findings:

> The abnormal constraints imposed by captivity result in a variety of bizarre behaviors of which female-female matings is but one. A far more common one is pseudocopulations between males of bisexual species, such as *Cnemidophorus tigris*. In my laboratory, the larger or healthier males 'rape' subordinates at will, albeit unsuccessfully, as insertion of the hemipenis requires 'willingness' on the part of the mate, even in male to female 'rapes.' This behavior is so common that the subordinates become emaciated and would die from perpetual harassment, if the 'sexual offender' were not isolated. The implications are similar to those proposed by Fitzgerald and Crews, but it would be premature at best to propose that this abnormal courtship behavior is essential for successful reproduction in *C. tigris*.

The parallel suggested for Crews and Fitzgerald's observation, read literally, seems to add to their article. But of course the abnormal-

18. Crews comments in the margin of my paper, where I quote Cole's comment on the geckos living in houses, that this "is where *geckos* live naturally," The comment supports my point about the rhetorical importance of the natural/artificial distinction.

ity of this behavior implies that the mounting behavior Crews and Fitzgerald have seen is also abnormal, and that they should learn, as competent keepers of these lizards, to guard against it. The understatement of the last sentence ("it would be premature at best to propose" that male–male matings could be essential to reproduction) is part of its heavy irony. Cuellar clearly expects that this dismissal will close the issue.

Similarly, Cole and Townsend present themselves in their conclusion as cautiously undecided, lacking evidence, and open to new ideas.[19]

> It would be interesting to understand the cause(s) and function(s) of malelike copulatory behaviour among female lizards, but few data specifically and positively pertain to these points.

They leave the suggestion of alternative interpretations to references to other *Cnemidophorus* workers, reinforcing the sense that Crews and Fitzgerald are isolated in the research community. They do not definitely claim that the behavior is the result of captivity.

> Regardless of the interesting ramifications of this behaviour, there is no evidence that homosexual activities normally are involved in the reproduction of unisexual species of lizards.

Part of the rhetoric of Cole and Townsend's article is in its not making a counterclaim; by avoiding such involvement they further suggest that Crews and Fitzgerald have made their claim prematurely. Cole and Townsend do not say that they are reinterpreting the observations, but deny that the observations have been established as meaningful. The conclusion strikes a note of openness and caution but actually moves toward closure on the debate, implying that there is no basis for a controversy.

But the Cole and Townsend article ends with a postscript that makes an explicit attack on the competence of Crews's group as observers of behavior.

19. Crews comments "I don't read it this way." But this is because he sees the statement with a detailed knowledge of the context of the controversy, in which the statements that there are "few data" and "no evidence" imply dismissals of his group's work, not openness to further research. A reader coming to this article out of the controversial context might interpret the authors' stance as one of cautious indecision.

> A new paper (Gustafson and Crews 1981) that we received after completing this manuscript further confuses knowledge of the sociobiology of unisexual lizards because it erroneously relies on lifting of the hands during basking as an indication of submissiveness in *C. uniparens*. Such basking behavior in *C. uniparens* should not be confused with the arm or hand waving that appears to be a signal in other species, as in *Cnemidophorus lemniscatus*.

Again, as with the quotations of claims and the unusually detailed description of methods, such a direct and personal criticism would seem very odd except in the context of a controversy. The language of a postscript can apparently be more explicitly critical than that of the main body of the article, where Cole and Townsend only say that their results contradict those of Crews and Fitzgerald. As the personal notes in the methods section are unusual because they focus attention on the authors' competence, the personal note in this postscript is unusual because it focuses attention on the incompetence of the researchers they are criticizing.

As in the main controversy over the meaning of the mounting behavior, the reinterpretation depends on whether the behavior observed—the lifting of a front leg—is parallel to basking in the same species (in which case it has no meaning for reproductive behavior) or whether it is parallel to the narrative of submissiveness in another species (in which case it can be used as a signal of courtship). Behavioral terms similar to those in dispute here also figured in the construction of the behavior and in the CGT and MWBC articles. Cole and Townsend, instead of showing that what they consider to be a misinterpretation would bias the results of Gustafson and Crews, need only point out the apparent error to taint the whole research project. Crews's project is placed in the narrative of the whole field, as an obstruction that "further confuses knowledge of the sociobiology of unisexual lizards." The sweeping nature of this criticism suggests that narratives are arranged in a hierarchy of inductive argument, so that if a researcher can be shown to be wrong in an interpretation of an animal's action, then the study and the project of which the observation is a part both crumble. A researcher who cannot tell basking from handwaving has not just made a mistake, he or she is incompetent and misleading.

CGT end their defense with a move toward closure much like that at the end of the Cole and Townsend article. They refer to a narrative of the whole ongoing project, summarizing current knowledge such that their position represents fact and the other position represents hypothesis.

> We cannot yet say whether the *presence* or *absence* of male-like sexual behavior in captive populations of unisexual lizards is abnormal. However, the following observations are not subject to controversy and still require explanations: . . .

Both CGT and Cole and Townsend seek to remove certain issues from controversy to bring the issue, or at least part of it, to closure. Both strike the cautious note of saying that not enough is known, but Cole and Townsend take this lack of knowledge as a disproof, while CGT take it as a call for further research.

The MWBC article prepares for closure by turning the tables, asserting that the authors have put forth the experimental evidence and that they are involved in testing hypotheses, so that it is their behavior, rather than that of their critics, that is properly cautious:

> Gustafson and Crews (1981) have demonstrated experimentally that the presence and behaviour of cage-mates causes captive *C. uniparens* to produce more clutches of eggs. An understanding of the obviously complex social biology of unisexual *Cnemidophorus* will be advanced only by rigorous testing of hypotheses.

Only if one has the project of Crews's group in mind do these sentences follow one another. MWBC assert, after all the reasons for disregarding other research as flawed, one assertion that is supported *experimentally.* Of course no researcher would argue with the need for testing of hypotheses in general. But researchers might argue with the assertion here that the Gustafson and Crews article supports such a hypothesis in need of further rigorous testing.

The MWBC discussion ends with a closure move very similar to that of the earlier rebuttal, with the authors' position defined as the cautious one supported by data, and supportive of further research.

> Until these data [on pseudocopulation in wild populations] are collected, the only hypothesis that is supported by experimental tests with captive individuals (Gustafson and Crews 1981; Crews 1982) is that pseudocopulatory behavior is adaptive because it enhances reproductive potential.

This closure is an attempt to reinterpret the narrative of the whole field, to present their own project as a starting point rather than as a digression. The key citations for the project are now somewhat later articles than that of Crews and Fitzgerald; they focus attention on articles that have a more complex experimental design, shaped in the course of the controversy. The language of hypothesis and experiment is claimed only for this project; it is the others who are indulging in speculation.[20]

An End to Polemics?

Textually, this persistence of two inconsistent accounts could go on for ever; either side can put the narratives of the other in new contexts for ironic reinterpretation. Practically, in terms of funding and publication, either Crews continues his research, or he doesn't. The issue will be settled, not by something the lizards do, but by the dynamics of one part of the scientific community. That this particular controversy is not closed can be seen in the papers from a symposium on the *Cnemidophorus* held by the American Society of Ichthyologists and Herpetologists in 1984, which Crews said "promised to be a modern version of the shoot-out at the OK corral." This seems not to have been the dramatic occasion that the tone of the articles I have presented so far might suggest. But we can see signs of how other researchers were responding to the continuing work of Crews's lab in both Cuellar's paper and in the comments of referees who reviewed Crews and Moore's paper before its publication.

Cuellar's response seems to be to continue on his own line of research, continuing his doubts about the relation of the observed behavior to reproduction, but answering Crews's publications only where they directly criticize his methodology. Cuellar was not able to attend the symposium, but submitted a paper for the conference proceedings, "Further Aspects of Competition and Some Life History Traits of Coexisting Parthenogenetic and Bisexual Whiptail Lizards," As the title suggests, most of the article is devoted to issues unrelated to Crews's claim, and he responds to Crews, again, only in a postscript. Cuellar quotes the passages criticizing his observations that I have quoted, starting with the line, "Preconceptions, however, guide perception, and one does not very often see what one is [not] looking for." His tone in response to the response to his criticism involving bite marks is apparently mild.

20. Again there are parallels between these closings and those in the "Technical Comments" on Cuellar's 1977 *Science* article. Each text there acknowledges the existence of a controversy, and then concludes with a suggestion for closing it, and each of these suggestions identifies the author's own positions with facts, objectivity, openness, or usefulness for further research.

> These are legitimage points raised by Crews et al., for indeed I had not previously documented the frequency of these bite marks, and had assumed they would stand out conspicuously during routine examination of freshly-collected live animals.

But there are two criticisms of Crews' response implied here: it is suggested that these marks should be noticed as a matter of competent collecting, and also that Crews's study—looking for the marks on *preserved,* rather than freshly collected specimens—would not provide relevant evidence. Cuellar then goes on to present extensive data about bite marks in the field, data Crews's group had said were lacking. As he himself points out, such data would not have been tallied, much less published, were it not for controversy that made them suddenly become relevant:

> Prompted by the report of pseudocopulations in the laboratory by Crews et al. (1981), and by the challenge by Crews et al. (1983) to document my field observations, I have since recorded the location and extent of the marks and correlated their occurrence with reproductive condition in samples captured and released during three years from 1982 to 1984 (Table 5).

In interpreting these data, Cuellar makes a point of his giving every possible benefit of the doubt to Crews's case. His conclusion is that while the marks occur, they do not occur in such a pattern as to support the claim that pseudocopulation facilitiates reproduction. He continues to quote publications from Crews's group extensively and always ironically. For instance, he quotes the comment in MWBC about the difficulty of observing these shy lizards, and then quotes his own work describing *C. uniparens* as relatively easy to observe. He ends by turning their criticism of him back on them:

> In fact, the extensive laboratory documentation of such behavior strongly suggests it is common in the field, but as Crews et al (1983) appropriately note "perceptions being subjective are not readily changed by argument and riposte."

This may seem to grant Crews and Fitzgerald's original claim, but on what is now the key issue, the relevance of this behavior to reproduction, he remains unconvinced. There is a kind of closure here: he has decided, as has Crews's group, that there is no chance of persuading the other side, that there is a limit to the amount of back and forth criticism that is worthwhile, and that resolution will come only by persuading those *Cnemidophorus* workers who haven't been involved in the controversy.

Crews and Moore's paper for the same ASIH conference, "Reproductive Psychobiology of Parthenogenetic Whiptail Lizards," opens with a reference to Crews and Fitzgerald (1980) and then reviews the work on pseudocopulation up to 1985. The reviewers of this paper, who are *Cnemidophorus* workers uninvolved in the controversy, are generally enthusiastic about Crews and Moore's work, but less enthusiastic about their rhetoric. One begins: "Overall, this is a well done and important paper presenting additional information on an inherently interesting topic. As presented, however, it will continue to foster controversy and rabid-dog type criticism. Some of the reasons for criticism are valid." This reviewer thinks that the whole issue of whether pseudocopulation occurs in the field—the issue that led to the back and forth exchanges on bite marks and on the shyness of the lizards—leads researchers away from more important issues: "The important point is that whether or not the unisexual species do it in the field, they certainly do it in the lab and this offers a unique opportunity to observe a 'male' behavior in the absence of the heretofore assumed payoff, insemination." This reviewer would resolve the controversy by redefining the context. It is an important step toward a resolution that would acknowledge Crews's interpretation while leaving open its relevance to behavior in the field (this is analogous to the kind of resolution Martin Rudwick sees in *The Great Devonian Controversy*). But for Crews's evolutionary argument, it would seem that the occurrence of the behavior in the field does matter, and he is not ready to abandon that part of his argument so easily.

All the reviewers question the elaborate negative evidence for the failure to see the behavior in the wild, but they do not go on to question the "naturalness" of the behavior itself. One of the reviewers is especially dubious about the sort of redefinition of the claim of Crews and Fitzgerald that I have discussed:

> The authors request here an end to polemics (good idea!), but then appear silly in defending a poorly stated conclusion in the 1980 paper. To most readers, "successful reproduction" means "producing any offspring." The authors claim they meant to say that degree of reproductive success (i.e. number of offspring) is related to

pseudomale behavior. It seems time to bury the hatchet . . . who's going to go first?

The reviewers' comments suggest that a controversy may sometimes be resolved, not when new evidence comes in to settle it, but when everyone else gets tired of it, and finds ways of either using or getting around the new claims. At this stage, as one reviewer points out, it is still necessary to follow the story of the controversy from the beginning. This reviewer suggests that the volume include an essay by "one of the critics." The general reader "might be a bit confused without such a chapter." (The reviewer even suggests that the critics and Crews and Moore collaborate on a review, though this seems not to have happened.) But the controversy is approaching the point when these arguments will no longer matter. When a controversy is closed, as Collins has pointed out in his study of research on gravity waves in *Changing Order,* all the social processes will be forgotten. The construction of narratives and their ironic reinterpretation will no longer be an issue. There will be only the story of the lizards—not necessarily the same as the first story presented by either side. And the stories of the studies, projects, and the field will be subsumed into the one exemplary story of the progress of science.

One place one might expect to find this exemplary story would be in popularizations. As it happens, both Cole and Crews have written *Scientific American* articles on *Cnemidophorus.* If these articles are any indication, research continues without any generally accepted view of *Cnemidophorus* behavior because the various groups are able to pursue two completely separate lines of research. Cole does not mention Crews at all in his *Scientific American* article "Unisexual Lizards" (nor does he mention Cuellar). At the end of his article, he presents a story about the study and significance of the physiology of parthenogenetic lizards:

> Today interested workers find themselves in a position not only to ask new questions and design new experiments but to utilize these specialized organisms in ways that would not have been imagined a few years ago. Among the possibilities that come to mind are gaining a better understanding of the role of sperm in fertilization, clarifying how it is that some animals are quite successful with multiple copies of genes whereas others are not, studying switching mechanisms in embryonic development, producing cloned animals of known genetic composition for biological experimentation, and even inducing cloning in normally bisexual species to increase pro-

> ductivity in animal husbandry. If unisexual reptiles contribute to progress in any or all of these areas, it will have begun with a few startling observations concerning a form of wildlife that practically no one considered significant.

Crews, in his own *Scientific American* article, "Courtship in Unisexual Lizards: A Model for Brain Evolution," cites Cole's article in his paragraph on the various species of the genus and their chromosome makeup. But he makes a more important reference in his introduction to a much older *Scientific American* article by his mentor Daniel Lehrman. Crews's story, as summarized in his introduction, is really about circuits in the brain controlling behavior; the lizards are interesting as a natural experiment showing how these controls work.

> The brain, which controls mating behavior in males and females, not only has adapted to a new set of stimuli in this species but has also mediated a switch to females of behavioral patterns that are normally associated with males. This reinforces the observation that the brain is equipped with neural circuits for both male and female behavioral repertoires, regardless of biological sex. By investigating the manner in which that has come about, using unisexual lizards as my model, I have gained insight into the ability of the brain to adjust to changing conditions during the course of evolution.

These underlying differences in the narrative of the discipline into which they insert their studies make for quite different approaches throughout the articles. For instance, they both discuss the descent of the unisexual species from hybrids of other species, but they cite different evidence and different researchers on different species. Cole cites traditional descriptive work done in the 1960s by two important *Cnemidophorus* workers:

> Lowe and Wright found that in such pertinent attributes as color, color pattern, scale shape, chromosomes and preferred habitats the character of *C. neomexicanus* appears to be that of a first-generation hybrid produced by the mating of *C. inornatus* and *C. tigris*.

Crews cites more recent work that is far from the traditional methods of natural history:

> By comparing the DNA sequences of various whiptail lizards, Llewellyn D. Densmore III, Craig. C. Mortiz and Wesley M. Brown of

> the University of Michigan were able to determine that the maternal ancester of *C. uniparens* is the bisexual species *C. inornatus.*

There is no conflict between these two findings, but each author uses an approach to the determination of descent that is most appropriate to the methods of the rest of the article.

Since the *Scientific American* editors published both accounts, they must consider both Cole's line of research and Crews's to be of general interest to the public, and to be generally accepted by the *Cnemidophorus* community. There is no confrontation between them, and indeed, neither article mentions any controversy. I am not suggesting that either Cole or Crews is hiding something in these articles, just because they don't dwell on the controversy that interests me. They are following a convention of popularizations, by which writers usually present the current consensus on any topic, not still-controversial views by individual researchers. What is it about popularizations that eliminates just those features that tend to show the social construction of science? In the next chapter I shall look more closely at the construction of popularizations.

Whenever I have presented papers on this controversy, it has struck audiences as funny that scientists would argue with such heat about whether a lizard's lifting of a leg is hand-waving or basking, or about whether there are bite marks on a lot of preserved specimens of dead lizards. If it were only the technical details were at stake, the argument would indeed be trivial, and probably wouldn't even interest other *Cnemidophorus* workers. But such technical details have their place within narrative of studies, that themselves have their place within a larger narratives of the field; Crews, for instance, sees implications for the important question of the hormonal control of human sexuality. In fact, there are few controversies in biology that do not have broader implications. The implications are brought out most clearly, not in controversies in the core set, or in typical popularizations like these, but in a few persistent controversies that take place in the public forum. These public controversies require writers to address a different audience and use different techniques, but one still sees the basic strategies of construction and ironic reinterpretation of narratives. In chapter 6 I shall examine the strategies in one such public controversy, in which participants display the larger implications of technical details, in a study of response to E. O. Wilsons's *Sociobiology*.

Chapter Five

The Social Construction of Popular Science: The Narrative of Science and the Narrative of Nature

My stance in the studies so far is to assume that many readers will be surprised by the view that science is constructed in social processes of claims and negotiations, carried out in revisions of articles and proposals and in ironic reinterpretations in controversies. This stance assumes that many readers, especially nonscientists, will start with a different view of the work of science from that which I am proposing, a view that sees the main work of science as passively observing naturally occurring facts. But if people do hold this view of science, where would it come from? And why would anyone come to think of scientific texts as just conveying information? I shall argue that even very sophisticated popularizations tend to promote a view of science that focuses on the objects of study rather than on the disciplinary procedures by which they are studied.

Those who have studied popularizations have generally agreed that articles for the general public and articles for scientific specialists are strikingly different, but there is tendency to take either articles for popularizations or specialist articles as primary and dismiss the other form as a distortion. Either the popular article is seen as watering down the difficult truths of the professional version, giving the false impression of easy comprehension, or the professional version is seen as complicating the simple truths of the popular version unnecessarily, using jargon and technical details to exclude untrained readers. These two accounts are evident, for instance, in the responses to a striking experiment conducted in 1971 by F. J. Ingelfinger, the editor of the *New England Journal of Medicine*. Exasperated with immunology articles so difficult that only other immunological researchers could read them, Ingelfinger had one rewritten by Barbara Culliton, a journalist on the staff of *Science*, and published both versions. Culliton kept all the information in the original, but reorganized the article, rewrote the sentences, and included some definitions of terms in appositives, so that any practicing physician could read it. The editor received a number of letters saying he had proved that even difficult topics could be made accessible to a wider audience with some attention to organization and clear writing. But he also received some letters from immunologists saying that they found the revised version harder to read; it was as if the housekeeper had come in and nothing was where they were used to finding it. Both groups of letter writers thought that the experiment showed that there was a right way and a wrong way of writing immunology; they disagreed only about which was which.

I shall use the approach to narrative introduced in chapter 4 to argue that popularizations and scientific articles present two views of what a scientist does, two views that are incompatible but that both play a part in creating the cultural authority of science. I shall look at the ways the narratives are constructed in articles in some articles on evolutionary biology in *Science* and *Evolution* and in articles by the same authors on the same topics for more popular journals, *Scientific American* and *New Scientist*. Textual differences in narrative structure, in syntax, and in vocabulary can help define two contrasting views of science. The professional articles create what I call a *narrative of science;* they follow the argument of the scientist, arrange time into a parallel series of simultaneous events all supporting their claim, and emphasize in their syntax and vocabulary the conceptual structure of the discipline. The popularizing articles, on the other hand, present a sequential *narrative of nature* in which the plant or animal, not the scientific activity, is the subject, the narrative is chronological, and the syntax and vocabulary emphasize the externality of nature to scientific practices.[1]

1. The two categories I use may be compared with those in several recent sociological studies of scientific discourse. In *Opening Pandora's Box,* Nigel Gilbert and Michael Mulkay distinguish between two ways scientists account for their work. In the empiricist repertoire of formal scientific discourse, actions are explained in terms of purely scientific factors, whereas in the contingent repertoire, excluded from formal scientific discourse, actions are explained by other social and personal factors (p. 40). But in these terms, both the narratives I describe use the empiricist repertoire. Both the scientist and the public have an interest in treating the facts of science as something apart from contingent processes. The narrative of science accomplishes this separation by certifying the acceptability of the methods and concepts used; the narrative of nature accomplishes this separation by focusing on the object studied and excluding the conditions of study.

I define these two narratives through two kinds of comparisions. First I illustrate the kinds of textual features relevant to these two narratives by comparing published professional and popular articles by the same authors on the same research, looking especially at differences in titles, abstracts, introductions, organization, and illustrations, features that guide the reader in constructing a narrative. Then I compare the manuscripts of the popular science articles by scientists to the published versions as edited and extensively rewritten by the editors of popular journals. In this comparison I go from large-scale changes to small: from changes in the overall organization of the text to changes in the syntax of sentences and of individual words. The negotiations between authors, who try to write a narrative of science, and editors, who want more of a narrative of nature, are where these two views of science meet. Finally I shall briefly compare popularizations of these authors' studies in several different publications for several different audiences.

The differences between these discourses have implications for the study of the public understanding of science. Many studies of popularization treat science as information that is merely communicated to nonscientists in more or less efficient language. As in the previous chapters, my approach is based on the assumption that science is embodied in language, so the translation of one form of words into another changes the meaning in some way. Even when two articles seem to be about the same research, it may turn out that one is about garter snakes and the other about isolation of a pheromone. One consequence of this assumption is that we should not expect the writers or readers of either narrative to enter easily into the other.[2] If

---

The distinction between the narrative of science and the narrative of nature also parallels the levels of "externality" analyzed by Steve Woolgar ("Discovery"), and in a different way, Trevor Pinch ("Towards an Analysis of Scientific Observation"); these scales are discussed in chapter 3. Pinch shows that scientific texts are chararacterized by a tendency to claim the greatest externality possible. I shall argue that popular texts, on the other hand, assume the externality of all scientific findings, and omit whenever possible any suggestion of scientific artifice.

Still another distinction that could be usefully compared to mine is Michael Lynch's analysis of "talk about science" and "talking science" in *Art and Artifact in Laboratory Science*.

2. This comment does not mean to imply that there are not many biologists publishing important specialized research who also publish popular science articles. Such scientists, I would argue, can handle both the narrative of science and the narrative of nature; they do not necessarily make the two narratives indistinguishable. For a lively presentation of the changes necessary in popularization, by one of the masters of the genre, see J. B. S. Haldane, "How to Write a Popular Scientific Article," now collected

words embody science, then both sides, the professionals and the public, have a stake in the form of words they use. As we saw in the introduction, many studies have shown that the narrative of science is part of what maintains the scientist's sense of objectivity and cumulative progress and the definition of the discipline. The popularizers have a stake in popular language, in the narrative of nature, as well. Scientists interested in the public understanding of science should consider, not just attitudes toward science or how much scientific knowledge the public has, but how the public interprets scientific activity.

The Narrative of Nature and the Narrative of Science

I shall base my descriptions of the narrative of nature on two articles in *Scientific American* and one in *New Scientist*. The publication process for these journals is different from that of the professional journals with their competitive peer review; both journals commission articles by authors recommended to them and edit the articles to suit their audiences. *Scientific American* is an American monthly with a general audience; many of its readers have some scientific or technical training. It publishes rather long articles (authors are told to keep them to about 4,000 words), all of them by research scientists. *New Scientist*, a British weekly, has shorter articles (2,000–2,500 words) and a broader readership that includes many secondary school students. Gail Vines, one of the editors (in a letter pointing out that scientists' articles sometimes need to be edited to make them readable) notes that this readership is not exactly the general public:

> Our market research . . . tells us that half our readers have at least one A level in science. People working in science say they read it to

---

in *On Being the Right Size* (Oxford: Oxford University Press, 1985). Other well-known examples of scientist-popularizers include Julian Huxley, Peter Medawar, James Watson, Francis Crick, and Stephen Jay Gould; I would also think of Mark Ridley, Richard Dawkins, John Maynard Smith, and E. O. Wilson, in the areas at which I have been looking.

Robin Dunbar points out that the heavy editing I describe is not necessarily the rule; a recent article on his area of research was accepted with only the deletion of one sentence by *New Scientist*. But Dunbar has written often for *New Scientist*, and may be said to have internalized their style. The article focuses on the animals in the way I say is typical of the narrative of nature, but it contains much more about scientific debates than any of the articles studied in this paper.

> keep up with other fields. The maxim in the office is that the physicist should be able to understand the biology (and vice versa).

Many of its feature articles, like the one I shall study, are by researchers themselves, but other material is reported by staff journalists and freelance science writers. Both journals have extensive illustrations, usually photographs in *New Scientist* and elaborate and lovely paintings and color charts in *Scientific American.*

All three of the authors I am studying in this chapter had recently published a number of articles reporting their research in professional journals. I have chosen for comparison articles in *Science* and *Evolution. Science* is one of the two weekly general science journals that provide biologists with a prestigious outlet for rapid publication (see chapter 3). *Evolution* is one of the core journals of evolutionary biology; that is, it is a journal read regularly by nearly all specialists in a broad field.

Why do scientists write for the popular journals, when all the professional rewards are for articles in professional journals? Not for the money; the fee is small, considering the disproportionate amount of time such articles take to write (though one researcher I interviewed paraphrased Samuel Johnson's comment that no man but a blockhead ever wrote, except for money). They don't get rewarded with citations either; these journals are not usually places for first reports or findings, and they do not allow for extensive review or theoretical development. But there is clearly prestige within the research community attached to being asked to speak for one's field, and there is the chance to address a broad audience that includes many researchers and administrators in related fields who would not ordinarily read one's work in specialist journals. One of the authors I am studying tells his coauthor in a letter, "Remember that this article is as much an advertisement as it is informative." The writing of such advertisements is in many ways similar to the writing of the introductions to grant proposals; in both cases the researcher must put his or her work in its larger disciplinary context. But in popularizations there is a convention of presenting the representative nature of one's own work (and thus the author's appropriateness as spokesperson for the field), rather than stressing its uniqueness (and thus the author's worthiness for funding, in competition with others in the field). Although such articles may not directly advance the career of the individual writer, they are essential to the survival of the discipline, dependent as it is on public support for research. A 1985 report by the Royal Society on *The Public Understanding of Science* says, in bold type, "Our most direct and urgent message is for the scientists—learn to communicate with the public, be willing to do so, indeed consider it your duty to do so"(p. 24).

Why do readers read these journals? Advertisements for both journals suggest that readers are interested in technological developments, in scientific controversies, in the newness of ideas, in ideas with immediate practical implications. The journals offer, not only entertainment, but access to a kind of power. The articles make no attempt to draw the reader in, as scientific features in general-interest magazines and newspapers must do. A *Scientific American* article definitely takes as given the reader's curiousity about the topic, whether it is the sexual behavior of lizards or the opperation of zippers. *New Scientist* tries harder to be catchy, but still assumes a reader interested in the subject matter and not just in its current news value. None of the three articles I have studied tries to attract readers' attention by linking the topic to some popular debate or public interest. This is odd, because all of them can be presented in such a controversial context: one writes on ecology, another on the roles of hormones and of environment in controlling sexual behavior, and the third on sociobiology.[3]

I present three researchers, rather than just one, because, as the brief descriptions suggest, their research methods, and problems of popularization, vary in many ways. One effect of popularization is that they come out sounding rather the same: they are all presented as direct observers of nature in the natural history tradition. But in each case, they could also be seen as biological thinkers participating in debates over biological concepts and addressing various discipline-specific problems.

Lawrence Gilbert works on the problem of coevolution: how the evolutionary changes of two species in the same enviroment relate to each other. In 1981 he and Kathy Williams, then an undergraduate student working with him, published in *Science* an article reporting their studies on how the passion vine mimics butterfly eggs on its leaves and prevents the butterflies from laying real eggs on it. He was then asked to write an article on this topic for *Scientific American*. As

3. In Geoffrey Parker's article, "Sex Around the Cow-pats," the editor cut out the concluding comments on applications of sociobiological findings on sexual selection to man. Perhaps the section was cut, not because it is sensational or controversial or unrelated to the main point of the manuscript, but because the parallel of man with nature here makes sense only if one sees them both as following the same model of evolution. To see this one would have to focus on the concepts of the article rather than on the animals themselves.

we shall see, part of the popular fascination of his work is that it takes place in an exotic setting and deals with butterflies, which are beautiful, delicate, and perhaps more appealing to nonbiologist than, say, garter snakes or dung flies.

David Crews's work is also of interest to a broad range of readers, but is also, I think, open to misinterpretation by nonbiologists who see him as just a voyeur watching the sex lives of lizards and snakes. As we have seen in previous chapters, he studies the evolution of the systems that control reproduction, and of sexual behavior. He works through observation of the animals' behavior, examination of organs, and analyses of the substances in the animals' blood, and through comparisons of his findings to the findings of other researchers working with other species. But what comes across most strongly to a popular audience is simply the strangeness of the mating process in these species—not the parthenogenetic lizards we saw in the last chapter, but some garter snakes that live in the Arctic. He asked *Scientific American* to allow him to have as coauthor William Garstka, then a graduate student at Harvard, now an Assistant Professor at the University of Alabama. Garstka was first author of a *Science* article that provides the basis of much, but not all, of this popular article. Crews had written a previous article for *Scientific American* (and has written another one since; see chapter 4) so he was familiar with their expectations and editing techniques, and he warned Garstka in a letter, "I fully expect some changes, perhaps extensive."

Geoff Parker's work can also be seen in two quite different perspectives. Parker was asked in 1978 to write an article on his work for *New Scientist* after a series of nine articles in various journals between 1970 and 1975, in the course of which he presented a mathematical model for various aspects of the mating behavior of dung flies. The model has a purpose in a larger sociobiological controversy; it shows that the behavior he had observed accords with the assumption that certain behaviors are selected through evolution. Unlike Crews, Parker had had no experience with popularization before, and he found it took a considerable investment of time and rewriting to learn to write for this new audience.

In each case, nonbiologists have difficulty conceiving of the activities of these researchers the way the researchers themselves see their activities. One source of the difficulty is suggested by Ernst Mayr's comment on the popular reporting of biology.

> Discoveries are the symbol of science in the public mind. The discovery of a new fact is usually easily reportable, and thus the news

media also see science in terms of new discoveries . . . Yet to think of science as merely an accumulation of facts is very misleading. In biological science, and this is perhaps rather more true for evolutionary than for functional biology, most major progress has been made by the introduction of new concepts or the improvement of existing concepts. (*Growth of Biological Thought*, p. 23)

In each of the cases I am presenting, what is so difficult for the public to understand is a concept based on evolution: coevolution of populations, adaptation of control systems, and evolutionarily stable strategies. What makes these concepts so difficult, I will argue, is not that they are forbiddingly abstract, but that in order to see why they are useful concepts one must also see science as a set of socially defined disciplines in which there is conflict and change. The news media present science as an accumulation of facts, not just because such an accumulation is more easily reportable, but because the value of such an accumulation to the public is reassuringly certain.

The value of discipline-specific conceptual structures and of debates among scientists is not so easily seen. Thus the popular accounts of the researchers I study stress their discoveries. A reader of Gilbert's article in *Scientific American* will picture him walking though the jungle and discovering the struggle between the butterfly and the passion vine. The reader of *Science* will see him, if he is imagined at all, in his greenhouse or his office, manipulating nature and marshalling the textual support of other researchers. A reader of *Scientific American* will picture Crews with his crew in Manitoba, learning all he needs by watching the snakes, or cutting them open and seeing their structure, without any experimental or conceptual mediation. They may even think he discovered these creatures, though he cites the earlier workers who studied them. A reader of *Science* will picture him performing assays and making inferences from them. A reader of *New Scientist* will imagine that Parker's main activity is lying in fields watching dung flies, while another biologist, reading *Evolution,* would see his work as devising mathematical models.

In conversation, the authors describe the differences between their articles for professionals and their articles for popular audiences in terms of levels of information: they can't go into so much detail, or can't mention all the qualifications, for a general audience. This description is consistent with a view of science as an inductive activity in which facts lead to concepts. I argue that the information is there in each of the popular articles, but the connection to scientific activity is lost. In emphasizing the narratives, rather than the information, I try to show how different views of science can frame the same facts. These two views can be distinguished in an analysis of the sorts of textual features that most obviously distinguish the two kinds of article: the titles, abstracts, opening sentences, organizational devices, and illustrations.

Titles

Titles are crucial indicators because they show what the authors or editors think will arrest the eye of the typical reader skimming the title page, or will categorize the article correctly for a reader looking for it in an index.[4] Williams and Gilbert's *Science* title, "Insects as Selective Agents on Plant Vegetative Morphology: Egg Mimicry Reduces Egg Laying by Butterflies," states an index heading and a claim to be proved. The *Scientific American* article, "The Coevolution of a Butterfly and a Vine," states a topic to be described. In the same way, the Garstka and Crews *Science* article implies a claim: "Female Sex Pheromone in the Skin and Circulation of a Garter Snake." The title supplied by the *Scientific American* editor, "The Ecological Physiology of a Garter Snake," states a topic. In both cases, the titles imply two different time scales, one of the time of an experiment showing reduced egg laying or pheromone presence in the skin and circulation, the other of the millennia required for evolution of these populations, or the months required for reproduction of the snakes.

Parker's *Evolution* and *New Scientist* titles hardly seem to refer to the same topic. The *Evolution* title is bewilderingly precise:

> The Reproductive Behavior and the Nature of Sexual Selection in *Scatophaga stercoraria* L. (Diptera: Scatophagidae). IX. Spatial Distribution of Fertilization Rates and Evolution of Male Search Strategy Within the Reproductive Area.

This specifies an ethological topic (sexual behavior), an evolutionary topic (sexual selection), the scientific name of the species, and the relation of a quantitative finding (spatial distribution) to a behavioral feature (male search strategy). From the point of view of a biologist skimming the table of contents, the most important words in this title are the *ands* that link topics normally considered separately, while leaving the article to say just how they are linked. The *New Scientist* title links the two areas most intriguing to a general reader: "Sex Around the Cow-pats."

4. Charles Bazerman, "Physicists Reading Physics."

Abstracts

The abstracts of the articles by Gilbert and by Crews and Garstka (Parker's have no abstracts) confirm the emphasis on the work of the scientists and its importance to other scientists in the articles for professionals, the emphasis on nature and its fascinations in the popular articles. For instance, Williams and Gilbert's abstract for *Science* suggests the structure and argument of the article:

> Experiments show that *Heliconius* butterflies are less likely to oviposit on host plants that possess eggs or egg-like structures. The egg mimics are an unambiguous example of a plant trait evolved in response to a host-restricted group of insect herbivores.

The subject here is *experiments,* actions scientists perform; the structure of the article will follow the experiment/control comparisons. The key adjectives for showing the importance of these experiments are those that claim an *unambiguous* example of the relation in question, and with a *host-restricted group.* The *Scientific American* version changes the sentences into a narrative with the butterflies and vines as the subjects.

> *Heliconius* butterflies lay their eggs only on *Passiflora* vines. In defense the vines seem to have evolved fake eggs that make it look to the butterflies as if eggs have already been laid on them.

The key words here are the words that dramatize the situation: *in defense, fake eggs.* The narrative of this article will follow the time relation suggested in the summary: butterflies *lay,* vines have *evolved,* eggs *have already been laid.*

The abstract of the Garstka and Crews *Science* article also focuses on experiments, but on experiments that give a *since/then* structure to the article.

> Serums and extracts of tissues from the female garter snake (*Thamnophis sirtalis parietalis*) each act as a pheromone and elicit male courtship behavior when applied to the back of another male. Since pheromonal activity is present in the yolk and liver tissue of untreated females and can be induced with estrogen treatment in serums and livers of males, the pheromone may be associated with circulating yolk lipoprotein, vitellogenin.

The abstract makes an argument, in which the presence of the pheromone in yolk and its production in males given estrogen, taken together, suggest its association with vitellogenin. The *Scientific American* abstract is similar to that of Gilbert's article in emphasizing the unusual features of the story as natural history: the harshness of the environment, the precision of physiological control, and the spectacular appearance of the mating behavior:

> In order to survive in the harsh environment of western Canada the red-sided garter snake has evolved a precisely-timed cycle of physiology and behavior with several spectacular features.

Again the emphasis in the popular abstract is on the narrative of the animal.

Introductions

We can see by looking at opening sentences that the scientific articles by these authors are quite different while the popular articles make the three researchers' work sound similar. Like the titles, these openings are meant to attract the interest of a typical reader of the journal. For instance, Gilbert and Williams's *Science* article begins by outlining a problem for biologists:

> The idea of coevolution between insects and plants is attractive to biologists attempting to account for patterns of plant chemistry and the use of plants by insects. (1) However, it is difficult to demonstrate a causal connection between a plant characteristic and a particular selective agent [because most plants have so many plants and animals attacking them]. . . . One approach is to study plant groups that support only one or a few herbivore taxa.

Thus study of this plant, with only one major predator, presents "one approach" to the general evolutionary problem, an approach that is "attractive to biologists." Note that all three articles for professional journals have a citation after the first or second sentence; It is necessary to place the article immediately in the context of the literature.

The second sentence of Gilbert's *Scientific American* article also presents a problem, but it presents a problem for mankind, not for biologists:

> Perhaps the most significant category of ecological interactions in terms of the net transfer of energy in the global food web is the interactions between plants and animals.

The introduction goes on to discuss parasitic pollen carriers that both help and injure their hosts, so we are reminded immediately that this is a biological topic of great agricultural interest.

The opening of Garstka and Crews's *Science* article also stresses the way their findings fit into the existing scientific literature:

> In many vertebrates, urine, feces, and vaginal contents, as well as exocrine glandular products, function as sex attractants and serve to facilitate the location and recognition of mates (1). We now report an additional source for a vertebrate sex pheromone.

Stated this way, their findings would seem to be of interest mainly to other researchers on sex pheromones. The opening of their *Scientific American* article, on the other hand, stresses the problem the snake has in its extreme northern habitat, rather than the problem pheromone researchers have in locating pheromones:

> The red-sided garter snake (*Thamnophis sirtalis parietalis*) is found farther north than any other reptile in the Western Hemisphere. It ranges into Western Canada, where the winter temperature is often below −40° Celsius and the snowcover is often continuous from late September through May. . . . In the den the overwintering snakes undergo a set of profound physiological changes. Their blood becomes as thick as mayonnaise.

The reader is drawn into the article, not by a suggestion of the economic importance of garter snakes, but by the oddity of a snake in the Arctic.

Parker's professional and popular openings offer the widest contrast, for the whole first paragraph of his *Evolution* article is about concepts and approaches, whereas the first paragraph of his *New Scientist* article ranges over anecdotes of a number of species. The *Evolution* article begins:

> The present series of papers is aimed towards constructing a comprehensive model of sexual selection and its influence on reproductive strategy in the dungfly, *Scatophaga stercoraria*. The technique used links ecological and behavioral data obtained in the field with

> laboratory data on sperm competition, for which a model has already been developed (Parker, 1970a).

The appeal of this article to its biologist readers is the promise of a comprehensive mathematical model, and the link between findings of one method (field observations of behavior) and another (laboratory data on sperm competition). The *New Scientist* version begins:

> Why do peacocks sport outrageously resplendent plumage compared with their more conservative mates? Why do majestic red deer stags engage in ferocious combat with each other for possession of harems, risking severe injury from their spear-point antlers?

The reader here is drawn in by consistent anthropomorphizing of animal behavior: *sport, resplendent, conservative, majestic, harems, spear-point.*

Organization

One of the great popularizers of biology, J. B. S. Haldane, reminds scientists in "How to Write a Popular Scientific Article" that they will have to rearrange their statements for a popular audience, right down to the level of the phrases in a sentence.

> Try to make the order of phrases in your sentence correspond with the temporal or causal order of the facts with which you deal. Instead of 'Species change because of the survival of the fittest,' try 'The fittest members survive in each generation, and so a species changes.' Not that I like the phrase 'a species changes.' It would be better to say 'the average characters of the members of a species, such as weight or hair-length, change.' (P. 157)

Haldane's problem here, besides his usual conscientious wrestling with the qualifications necessary for precise statement, is how to reorder statements from the simultaneity of a research report to the chronology of what he calls "a coherent story." We see the same rearrangement in contrasting the professional articles we are studying to the popularizations.

Each of the professional journal articles constructs a different sort of narrative of similar materials, but these narratives all depend on rearranging a number of events into a simultaneous order of argument, Gilbert by comparisons, Crews and Garstka by since/then formulations, and Parker with the definitions of the parts of one formula. For example, Gilbert and Williams make the argument that the structures that look like eggs evolved to look like eggs, by linking findings in this sentence:

> That these structures have evolved specifically to mimic *Heliconius* eggs is indicated by the facts that (i) heliconiines are important defoliating agents of *Passiflora* (7); (ii) larvae of many *Heliconius* feed on congeneric eggs and larvae (6); and (iii) females exhibit great care in inspecting oviposition sites (6, 8).

The sentence compresses three separate narratives, concerning observations reported in 1975, 1977, and 1963, respectively, into one statement to serve as the starting point for the present research. Similarly, the article presents itself as one study among several parallel studies of coevolution; the notes cite supporting parallels in work on other butterfly species. At the end of the article there is a list like the opening list compressing all the successive experiments reported in the article into a simultaneous argument.

> We have demonstrated that (i) *Heliconius* females respond to the presence of eggs; (ii) this response has a strong visual basis (8) although chemical clues are not altogether excluded, and (iii), the response to egglike structures of *Passiflora* and to real eggs both reduces the possibility that real eggs will be laid after host discovery and increases the time required to oviposit.

Between the introductory summing up of the literature, and the closing summing up of the article's narratives, Gilbert and Williams's article is arranged in short narratives, each reporting a controlled experiment. Within each of these narratives, the sequence of events is arranged, not chronologically, but in a hierarchical order following the argument (figure 5.1). These narratives are dominated by the control group/experimental group comparisons, another kind of simultaneity. For the reporting of controlled experiments is framed to assure the reader that all the relevant conditions of one group (except for the experimental treatment) were experienced by the other; such reports are a way of reshaping time. Further narratives within narratives are contained in the notes setting forth materials and methods. The statement each narrative is to support comes at the end or near the end of the narrative, an order suggesting induction, the collecting of information leading to generalizations. Similarly the statement of the larger evolutionary importance of the *Heliconius/Passiflora* example comes near the end of the article, instead of at the beginning as in *Scientific American.*

Whereas the *Science* article is arranged by concepts divided hierarchically into small narratives of experiments, the *Scientific American* article is arranged in a large narrative following the activities of the butterfly and the vine.

> To answer this question one must understand three aspects of the interaction between the butterfly and the vine. The first aspect is how the female butterfly finds the host plant. The second is how the butterfly makes a choice between depositing its eggs or not depositing them. The third consists of the factors that affect the survival of the eggs and caterpillars after they are in place on the vine.

The experiments are still reported within this narrative, but they are subordinated to the chronology, instead of the chronology being subordinated to the argument.

Garstka and Crews's *Science* article also tries to make the events of research simultaneous, but their device is the since/then of result/cause argument, rather than the comparisons characteristic of Williams and Gilbert's controlled experiments. We have seen this structure, in which a series of details precedes a conclusion to which they seem to lead inescapably, in the abstract. It is also apparent in most of the paragraphs (I have emphasized the key words here).

> *Since* the female attractiveness pheromone of *Thamnophis* is present in the liver, *but not* in the fat bodies, of untreated females, *and since* estrogen treatment can induce the pheromone in the liver and serums of males, *we suggest* that the pheromone is either the lipoprotein vitellogenin or a lipid-rich part of that large molecule. *The finding* that yolk elicits male courtship when applied to males *further supports* this conclusion.

There are eight such since/then sentence structures in the seventeen short paragraphs of the article, most importantly in the abstract and the conclusion. The penultimate sentence of Garstka and Crews's article is similar in form to the sentence Gilbert and Williams use to bring their various findings to bear on one point:

> Because of the findings that (i) there is no sex or treatment difference in lipid staining within the epidermis, (ii) the epidermal lipid is trapped under a heavily keratinized layer, and (iii) lipid is present

on the outside of the skin, we suggest that the sequestering of the pheromone in *Thamnophis* is a consequence of an active process analogous to the ejection of poison in certain related snakes.

Another sort of atemporal arrangement in this article is the comparison to other species; findings on three other genera, from 1935, 1938, and 1980, are combined to show how the mechanism for this species could work. This sort of comparison is at the core of all Crews's work. It depends, not on the chronological sequence of research findings, but on the bringing together in one narrative of several separate sequences. So our focus as readers is neither on the organism, nor on the activities of the individual scientists, but on the conceptual structure of biology, the parallels between species and systems, in which these comparisons can be made.

In Crews and Garstka's *Scientific American* article, as in Gilbert's, the narratives of experiments are inserted into this larger framework of the narrative of the organism. So after the opening outline of the reproductive process before, during, and after mating, the article covers the isolation and action of the pheromone that attracts males, and the methods by which this pheromone reaches the skin. Then the article discusses the pheromone that makes the females unattractive when they have mated. Then it discusses hormonal relations after mating. Thus the experiments are seen as pieces fitting into a puzzle, the overall shape of which is given by the snakes' life cycle.

Parker's *Evolution* article also achieves simultaneity of a number of narratives, but its principle of organization is that of a mathematical formula. The formula describes what should be the end product of sexual selection; the males should behave so as to allow equal fertiliza-

Hierarchial order, as in the article

(numbers in parentheses are authors' notes; sentence numbers are added)

1.–In the first set of experiments, we examined the response of the butterflies to the presence of real eggs on *P. oerstedii,* , the host without mimics.

2. Host plants were available to the butterflies only during experiments, when females were presented with combinations of plant cuttings with and without eggs.

3. The cutting were of similar morphology, and *H. cydno* eggs were placed on tendrils near meristems where eggs are naturally laid.

4. Eggs laid in the course of each trial were immediately removed from the test plants.

5. Three types of *H. cydno* eggs were placed on the cuttings: bright yellow eggs, just as they appeared in the field; green eggs, which were eggs that had been tinted with food coloring and rinsed with distilled water to blend with the plants' coloring; and washed yellow eggs, which were yellow eggs washed with distilled water and which served as controls.

6. In each test of oviposition preference, the butterflies were offered four *P. oerstedii* cuttings; two had single eggs of one type and two had no eggs or had a single egg of a different type.

7. The cuttings were arranged at random with respect to one another and the buitterflies were allowed to oviposit until they lost interest in the plants.

8. Most trials lasted 1 to 2 hours and the butterflies laid eight to ten eggs per trial.

9. The oviposition behavior of *H. cydno* was consistent.

10. The butterflies, probably responding to a combination of olfactory and visual cues (11), usually noticed the host plants as soon as the plants were brought into the greenhouse.

11. While fluttering around a plant, they repeatedly tapped it with their antennae, then landed on the leaves to drum the cuticle with their forelegs, presumably using chemoreceptors to "taste" and further identify the plant (12).

12. They would then fly around the plant, tapping and searching for a satisfactory oviposition site, or reject the plant by flying away.

13. Often, when a butterfly noticed an egg or egg mimics, it would stop searching the plant and fly to some other part of the greenhouse.

14. Percent oviposition (ratio of number of eggs deposited to number of inspections) on plants with no eggs was significantly higher than on plants that had either a natural or washed yellow eggs present (Fig. 2, A and B) (13), indicating that the presence of a yellow *Heliconius* egg does indeed reduce oviposition on plants.

15. When eggs were laid on plants already bearing a yellow egg, they were usually placed several centimeters away on another part of the cutting.

The same experiment, reconstructed in chronological order

3. The researcher gathers similar cuttings (the gathering of the original stocks is described in a note).

5. a. Researcher gathers eggs and divides them into groups.
   b. Researcher tints green eggs.
   c. Researcher washes green and some yellow eggs.

3. The researcher places the eggs on the cuttings.

2. The researcher keeps the butterflies from plants, except during the experiment.

6. The researcher prepares groups of cuttings such as to offer alternatives.

7. The researcher presents the cuttings to the butterflies [and observes].

10. The butterflies notice the host plants.

11. The butterflies tap and drum the host plants.

12. The butterflies fly around the plant searching for a site, or fly away.

13. The butterflies stop searching and fly away if they see an egg or egg mimic.

4. The researcher removes any eggs laid.

8. The researcher stops the trial after 1 to 2 hours.

9. The researcher concludes that the behavior is consistent.

15. The researcher figures the significance of differences in ratios (using a method described in note 13).

16. The researcher presents this information on a graph.

Figure 5.1. Hierarchical and chronological order. Hierarchical order reset from "Insects as Selective Agents on Plant Vegetative Morphology: Egg Mimicry Reduces Egg Laying by Butterflies," by Kathy S. Williams and Lawrence E. Gilbert. *Science*, Vol. 212, 24 April 1981, p. 468. Copyright © 1981 by the American Association for the Advancement of Science.

tions for all localities around the cow pat. The first part of the article consists of a series of sections, each of which discusses a factor in the males' search, and each of which leads to a part of the mathematical model. For instance, one section compares the likelihood of successful mating for the male searching in the dung to rates for males searching on the grass, in relation to the total number of males searching. First Parker describes what the males do, then he describes what he and his wife did to observe the flies, and then he calculates "gain rates" for each strategy males could follow. This calculation fills that one slot in the formula. The formula removes the element of chronology from the flies' narrative, summing up the chances of all flies, and also removes the chronology from the scientist's narrative, telling us what he did only in its place in the development of the formula.

In the second part of the article, Parker compares the results predicted by the model to observations, and he attempts to account for the differences by introducing factors not included in the general model. Again the form is based on the relation between the formula and observation, but now, rather than derive parts of the formula from observations, he works out the whole formula and compares the results to his observations. Then he discusses the implications of the model for the evolution of sexual behavior, putting the implications last, as they are in the other two professional articles.

As with the *Scientific American* articles, Parker's *New Scientist* article must cover a much broader range of material than an article for a professional journal, summing up studies published over the course of years. Like the *Scientific American* articles, it organizes this material around the experience of the animal, in this case the male dung fly, first summarizing the mating process, then discussing the arrival of the males, the guarding by the males, the capture of the females as they leave, the behavior of the males after capture, and finally, the subject of the *Evolution* article, the strategies of searching. (This bit is discussed out of chronological order, perhaps because it requires an understanding of the other parts of the process.) In each section, Parker first calculates what the flies should do, then compares this to his observations. So the formal principle is the reverse of that of the *Evolution* article, in which behavior was given its narrative structure by the formula; here the formula is given its narrative structure by behavior.

Illustrations

The differences in the narratives of the articles for professionals and those for popular audiences are even more apparent in the illustrations than in the verbal texts. Because space is at a premium, most scientific journals discourage extensive photographs and figures. But the illustrations in a popular journal are a large part of the magazine's appeal to a casual reader; the illustrations in *Scientific American* are particularly lovely and eye-catching. They also contribute to the popular narrative's chronology, and to its focus on organisms rather than concepts.[5]

The Williams and Gilbert *Science* article has just two illustrations: a line drawing of the cuttings used in the experiment (figure 5.2) and a series of graphs comparing the rates of oviposition with various preparations of leaves. These show part of the preparation for the experiment, and summarize its results. The *Scientific American* article has beautiful drawings prepared from photographs provided by Gilbert, rather than stylized line drawings: detailed drawings of the butterfly and caterpillar, an elegant display of variations in leaf shape (figure 5.3). It also has three electron micrographs that, though they illustrate the rather subsidiary point that caterpillars can get stuck on spines, are the most dramatic of the illustrations (figure 5.4). The difference, then, is that *Scientific American* shows what these plants and animals look like; *Science* shows what Williams and Gilbert did. This difference is also apparent in the graphs used in *Scientific American* (also attractively done in colors; figure 5.5a); they illustrate all the graphs in the *Science* article except the one that shows the control group (figure 5.5b). This graph is unnecessary in the *Scientific American* presentation because it illustrates a feature of experimental design, not a feature of nature.

5. Nigel Gilbert and Michael Mulkay have discussed *Scientific American* illustrations in *Opening Pandora's Box*. They suggest that these illustrations give a physical reality to biologists' conceptions that are both more complex and less definite than the realistic picture would suggest. (A good example from the articles discussed here would be the cutaway drawing of a snake's skin in Crews and Garstka's *Scientific American* article.) I am making a somewhat different point about the apparent realism of the striking illustrations in the articles I am considering: not that they show details that are conjectural, but that they divert attention from the evolutionary argument to the appearance and stories of the particular animals and plants studied.

There is an excellent selection of articles on illustrations in Bruno Latour and Jocelyne Noblet, *Les 'Vues' de L'Esprit* (special issue of *Culture Technique* 14 Juin 1985). Michael Lynch discusses three scientific illustrations in detail and, in his notes, presents a very thorough review of studies of scientific illustration, in "Discipline and the Material Form of Images: An Analysis of Scientific Visibility," *Social Studies of Science* 15 (1985): 37–66. Lynch and Steve Woolgar have edited a special issue of *Human Studies* devoted to the theme, "Representation in Science"; my contribution is on the illustrations in *Sociobiology*, and other papers deal with a wide range of sciences (*Human Studies* 11 [1988]).

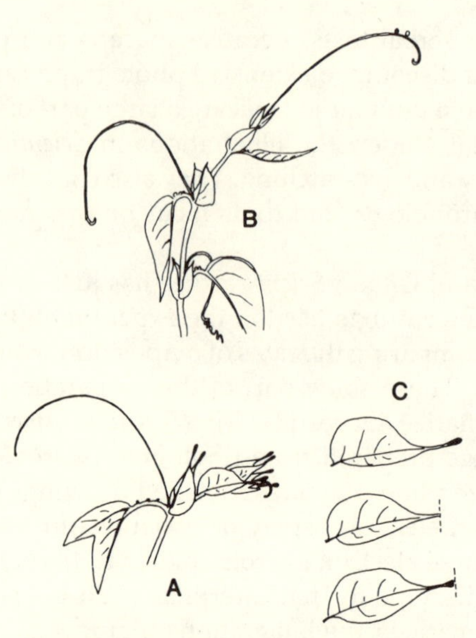

Fig. 1. ***Passiflora*** **cuttings used in experiment: (A)** ***Passiflora cyanea*****, showing display of egg mimics on stipule tips. (B)** ***Passiflora oerstedii*****, showing yellow egg (open circle) placed near green egg (closed circle) on tendril. (C) Enlarged view of** ***P. cyanea*** **stipules showing (top) unaltered stipule, (middle) stipule with egg mimic removed, and (bottom) stipule cut but retaining egg mimic for control.** ***Passiflora cyanea*** **stipules are 3 to 4 cm in length.**

Figure 5.2. A *Science* illustration. From "Insects as Selective Agents on Plant Vegetative Morphology: Egg Mimicry Reduces Egg Laying by Butterflies," by Kathy S. Williams and Lawrence E. Gilbert. *Science,* Vol. 212, 24 April 1981, p. 467. Copyright © 1981 by the American Association for the Advancement of Science.

The illustrations on every page of Crews and Garstka's *Scientific American* article also focus attention on the garter snakes themselves rather than on the biological point about the garter snakes. Dr. Crews's articles for professional journals often have graphs showing cycles of various hormones, but the *Science* article has, and needs, no illustrations. The *Scientific American* version, on the other hand, has a cover painting of the snakes, done by Ted Lodigansky, an artist commisioned by the journal, who worked from frozen specimens provided by Dr. Crews. The article is dominated by a color photograph, opposite the first page, of a mating ball, a large mass of male snakes. The next two pages of the article feature a series of drawings, done by a *Scientific American* artist from Dr. Crews's photographs, of the mating behavior

EGGLIKE YELLOW STRUCTURES appear on the three species of *Passiflora* (passion flower) vines shown in this painting. They mimic the yellow eggs of *Heliconius* butterflies that lay their eggs on the vines. The larvae of the butterfly then feed on the vine. At the left is a stem of the *Passiflora* species *P. cyanea;* the main modified egglike structures are the swollen ends of stipules: paired leaflike appendages. In the middle is a stem of the species *P. auriculata;* the main modified structures are nectar glands of the leaf stem. At the right is a stem of an undetermined species of passion-flower vine of northeastern Peru; the main modified structures are nectar glands of the leaf near the point where the leaf is attached to the leaf stem. In this species delayed expansion of stem-developing leaves keeps growth points hidden behind leaf displaying fake eggs. Growth points are vulnerable to being fed on by caterpillars that hatch out of real eggs.

Figure 5.3. A *Scientific American* illustration. Painting by Tom Prentiss, from "The Coevolution of a Butterfly and a Vine," by Lawrence E. Gilbert, *Scientific American*, August 1982, p. 111.

of garter snakes (figure 5.6). These four drawings outline the stages that, as I have suggested, provide the narrative for the article. The next two pages feature graphs of hormonal and gonadal cycles, illustrating the central findings of Dr. Crews's studies (figure 5.7b). Similar sorts of graphs in a later article in the journal *Hormones and Behavior* are such more stylized (figure 5.7a); *Scientific American* includes at each stage a little picture showing sperm in the testicles or showing little snakes growing in the eggs and then hatching. These certainly help the unbiological reader see what the stages mean, and they attract attention to what would otherwise be an off-putting graph. But they also help focus attention on the organism rather than on the concept of cycles, or on the measurement of hormonal levels and gonadal sizes that are the data reported here. The next two pages feature textbook-style illustrations of reproductive anatomy and some color micrographs by one of Crews's colleagues. These too give a sense that one is seeing the organism directly, rather than through the

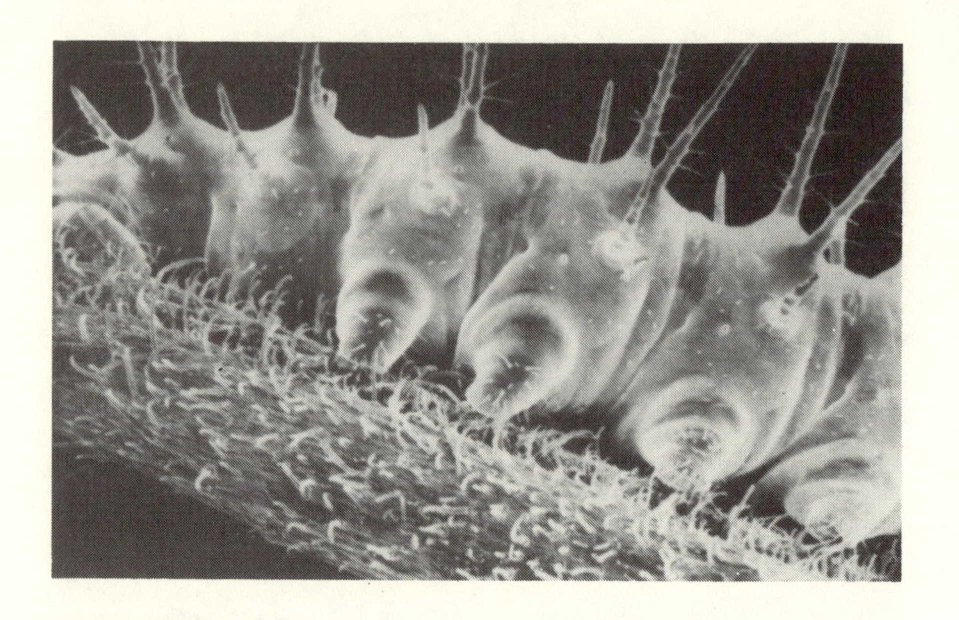

mediation of scientific theory and experiment. Finally the article illustrates the skin of the snake in a cutaway view like a radial tire advertisement, showing the hexagonal network of capillaries through which the pheromone reaches the skin (figure 5.8). It shows the path of the vitellogenin so clearly that the reader may wonder why Garstka

CAPTURED HELICONIUS CATERPILLAR, one of its prolegs hooked by a sharply pointed hair on the leaf stem of a *Passiflora* vine, is seen in this series of scanning electron micrographs showing another defensive measure of some species of the plant. At the top [A] the entrapment is not apparent; the hooked leg is the third from the right. In the middle [B] the tip of the plant hair has penetrated the surface of the proleg at the center of the micrograph. At the bottom [C] the tear that has been made in the proleg by the plant hair is visible just above the hook's point of entry. The vine in the micrographs is *P. adenopoda;* the caterpillar is *H. melpomene.*

Figure 5.4. An electron micrograph. These micrographs were arranged in a column when originally published in *Scientific American.* They have been rearranged here. From "The Coevolution of a Butterfly and a Vine," by Lawrence E. Gilbert, *Scientific American,* August 1982, p. 119. Copyright © 1982 by Scientific American, Inc. All rights reserved.

and Crews, or their predecessors, had any difficulty tracing it. Most of the *Science* article is devoted to the complex argument necessary to show that this is likely to be the pathway.

Parker's *Evolution* article contains three graphs illustrating the probability of capture of females (in the first part of the article) and comparing predicted and observed profiles for various search strategies (in the second part). The *New Scientist* article begins with a series of photographs of the mating process that function like the drawings illustrating Crews and Garstka's *Scientific American* article. It also includes three figures from his professional articles, with new captions. Considering the informal tone of the article, these presentations of mathematics come as a surprise. But we should note the way they isolate the mathematics from the rest of the article. And there is an interesting difference in the captions. The *Evolution* caption includes various adjustments and ends cautiously: "To emphasize that this profile must be regarded

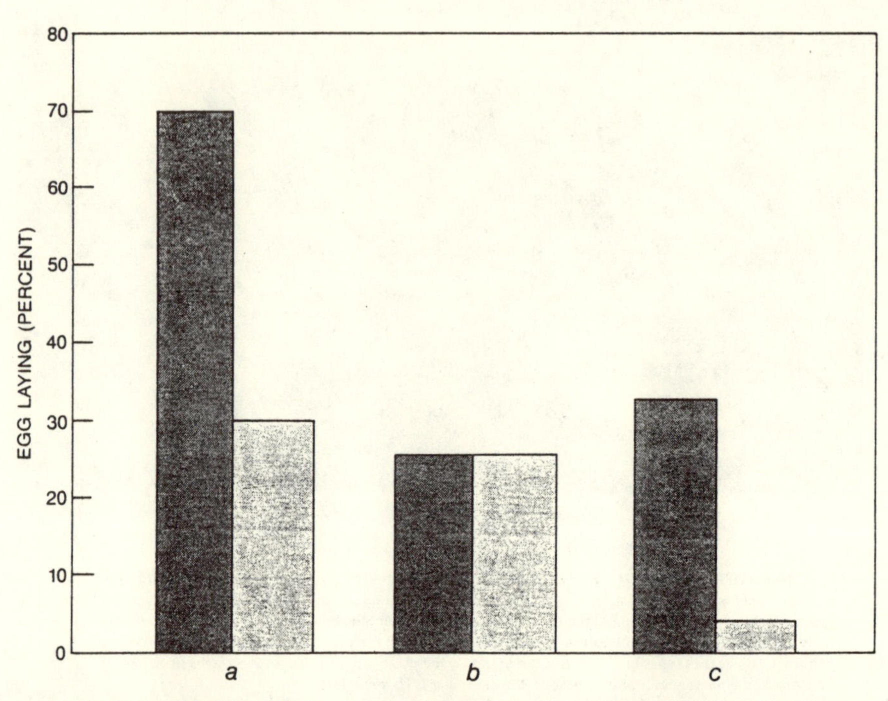

COLOR DISCRIMINATION in *Heliconius* females was demonstrated in a series of experiments with passion-flower vines. When *H. cydno* females were presented with a choice (*a*) between a vine bearing no eggs (*gray bar*) and a vine bearing an egg (*colored bar*), in a total of 217 inspections the butterflies selected the egg-free site 70 percent of the time. To determine whether color or chemical cues govern this behavior the butterflies' next choice (*b*) was between a vine bearing no eggs (*gray bar*) and a vine bearing an egg that had been dyed green (*colored bar*). In a total of 80 inspections the butterflies showed no greater preference for the egg-free site. Finally the butterflies were offered a choice (*c*) between a vine bearing a green egg (*gray bar*) and a vine bearing a normal yellow egg (*colored bar*). In 66 inspections the butterflies selected the site with the green egg more than 30 percent of the time and the site with the yellow egg less than 5 percent. Where the percentages in bars do not add up to 100 percent, the remaining fraction is accounted for by inspections in which the butterfly did not lay an egg.

Figure 5.5a. *Scientific American* graphs. The "colored" bars are those on the right. From "The Coevolution of a Butterfly and a Vine," by Lawrence E. Gilbert, *Scientific American*, August 1982, p. 114.

as approximate only, half the grid lines from each axis are omitted as compared with the expected profile." The *New Scientist* caption ends more confidently: "The fit between the two is encouraging." This example supports Parker's comment, in an interview, that popular versions are less cautious than professional versions; the two illustrations give the same graph, but the professional article emphasizes the differences between the model and nature whereas the popularization presents the model as a reflection of nature.

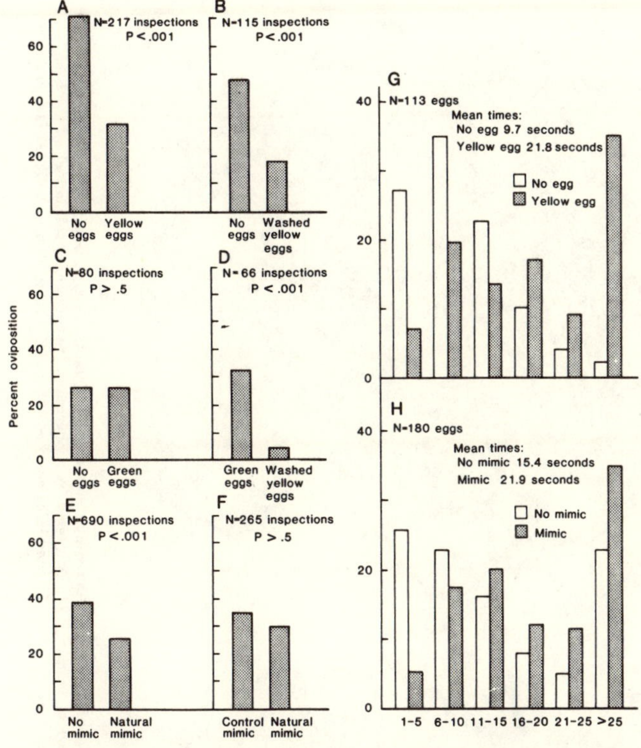

Fig. 2. Graphs which show oviposition response to various treatments. (A to D and G) Experiments were done with *P. oerstedii*. (E to G) Experiments were done with *P. cyanea*. In (G) and (H) the *x*-axis indicates seconds elapsed between recognition and oviposition. See text for details.

Figure 5.5b. *Science* graphs. From "Insects as Selective Agents on Plant Vegetative Morphology: Egg Mimicry Reduces Egg Laying by Butterflies," by Kathy S. Williams and Lawrence E. Gilbert. *Science*, Vol. 212, 24 April 1981, p. 468. Copyright © 1981 by the American Association for the Advancement of Science.

The Social Construction of the Narrative of Nature

The differences between the published texts of the popular and professional articles suggest two views of the activities of science. If we look at the revisions of the manuscripts of the popular articles by the editors of those articles, we can see these two views meet, and see how their differences are negotiated. I consider the changes made by the editors on three textual levels: (1) major changes of organization, (2) syntactical changes in many sentences, and (3) systematic changes in vocabulary. A nonscientist reader might see these changes as straightforward improvements that tighten the organization and make it easier to follow, bring out dramatic and memorable details, simplify syntax, and cut jargon. But the changes can also be seen as

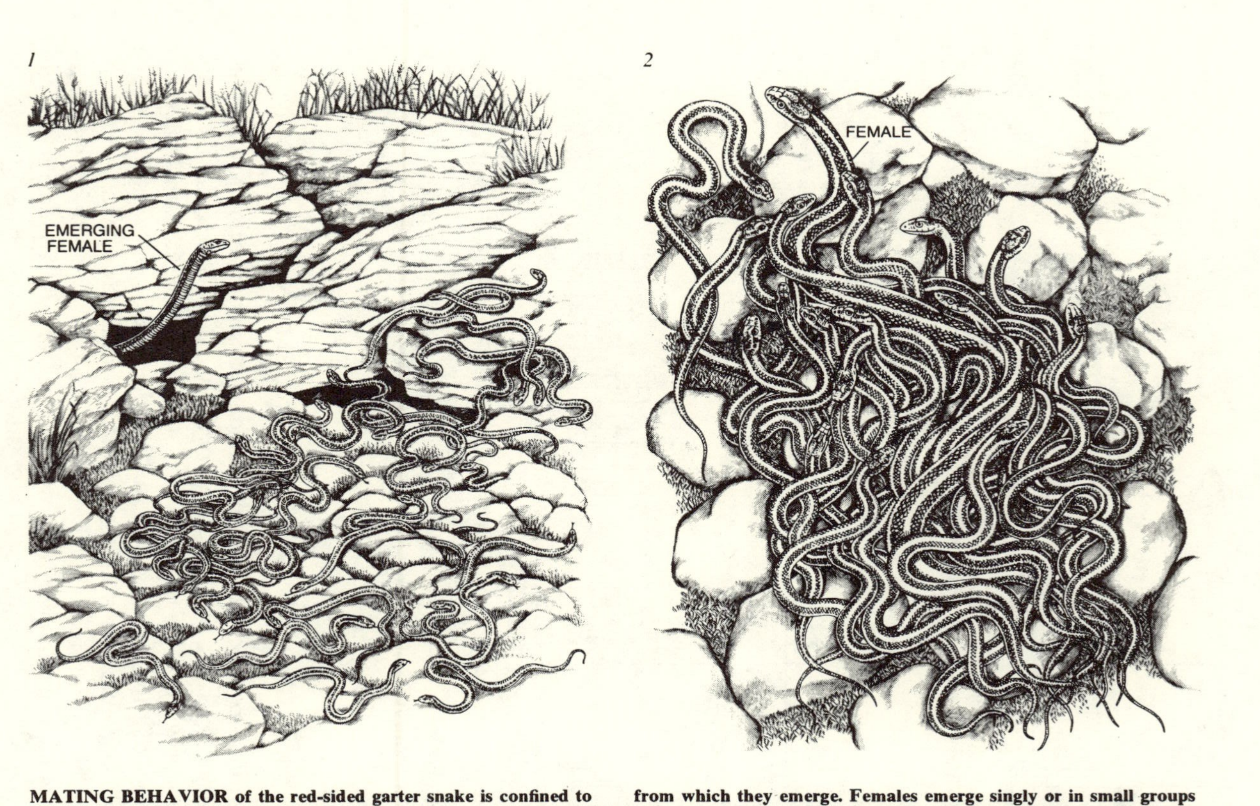

MATING BEHAVIOR of the red-sided garter snake is confined to a short, intense springtime breeding season. For a period of from three days to three weeks the males sun themselves near the den from which they emerge. Females emerge singly or in small groups (*1*). Attracted by a pheromone (a messenger substance) on the back of a female, as many as 100 males form a "mating ball" (*2*). One male

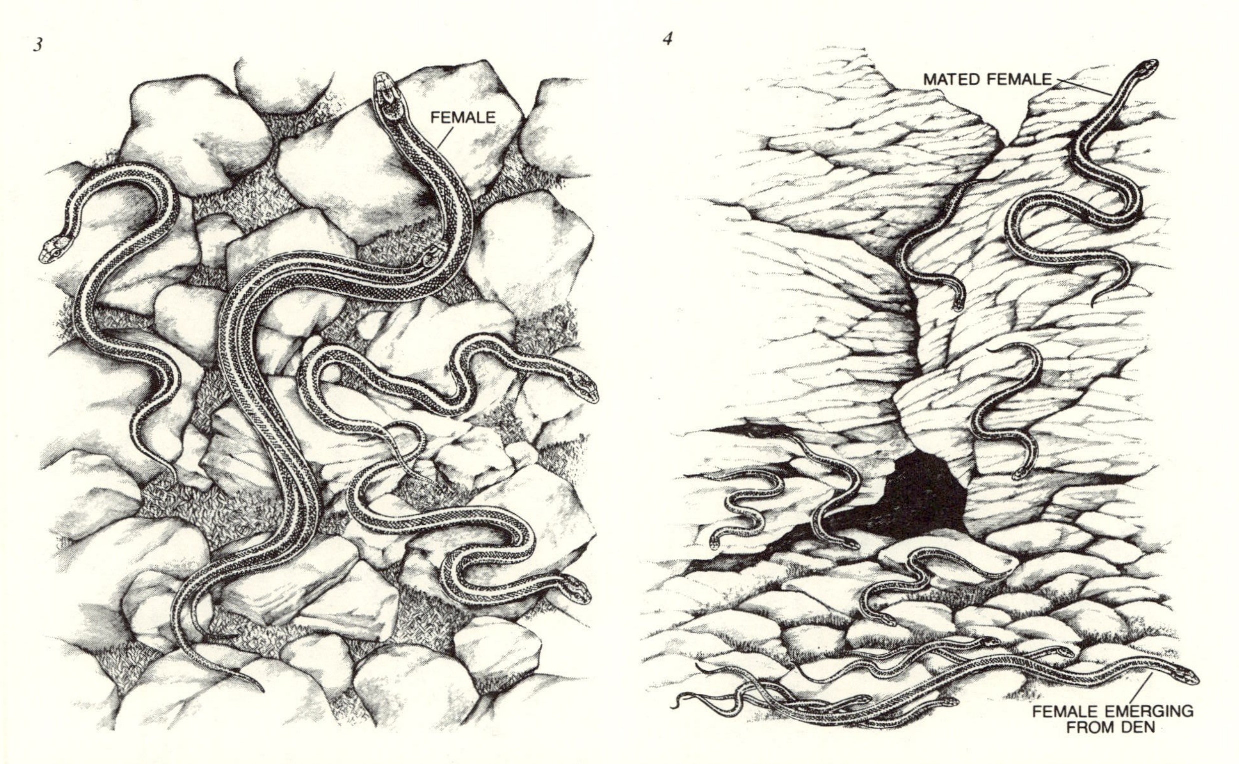

in the ball succeeds in mating with the female by inserting one of his two hemipenes into her cloaca (her urogenital opening). The other males immediately disperse (*3*). The mated female, rendered unattractive to males by a pheromone her mating partner conveys into her cloaca, immediately leaves the vicinity of the den. The males stay near the den to await the emergence of another unmated female (*4*).

Figure 5.6. A narrative in illustrations. Illustration by Patricia V. Wynne from "The Ecological Physiology of a Garter Snake," by David Crews and William Garstka, *Scientific American*, November 1982, pp. 160–61.

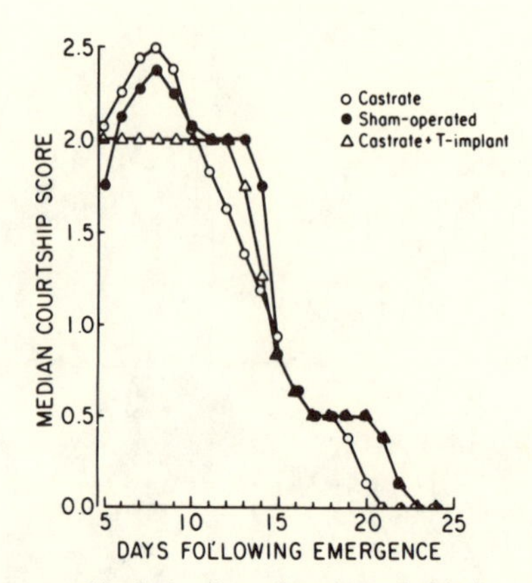

FIG. 1. Changes in courtship behavior of male red-sided garter snakes (*Thamnophis sirtalis parietalis*) on emergence from low-temperature dormancy. In nature following hibernation, or in the laboratory following low-temperature dormancy, courtship behavior initially is vigorous but then gradually declines in intensity; males will not exhibit courtship behavior again unless exposed to cold temperatures. Depicted here is the decline in courtship behavior in males that were castrated, castrated and given testosterone replacement therapy, or sham-operated in the fall prior to entering winter dormancy.

Figure 5.7a. A *Hormones and Behavior* graph. Illustration from "Hormonal Independence of Courtship Behavior in the Male Garter Snake," by David Crews et al., *Hormones and Behavior*, Vol. 18, p. 34. Copyright © 1984 by Academic Press, Inc.

subtly changing the message of the article, changing a narrative of science into a narrative of nature.[6]

6. A similar variability of views is evident in responses to my own paper. For instance, Gail Vines (an editor at *New Scientist*, though not the editor of Parker's article) points out that, in my effort to stress the changes involved in popularization, I go too far toward taking the scientific texts as primary:

> "I think you are too kind to the scientists. The style of academic journals creates a misleading air of "objectivity" which I think can be dangerous to both science and the public. I take your point that such articles also set a study in an explicit theoretical framework, but so do many good popularizations of science. Most weeks in *New Scientist* at least one article will be "theory-led." I wonder about the generality of your observations.
>
> Popularizations of science often do start with "nature" but I don't see how one can make a physicist understand the concept of sexual selection, say, without a few good examples of the phenomena that are, arguably, a result of the process."

Dr. Vines may well be right in pointing out the ways these articles are not typical. I have dealt with a wider range of popularizations in two other papers, "Making a Discovery" and "Reporting Genetic Fingerprints." I discuss the issue of authority she raises in my concluding chapter.

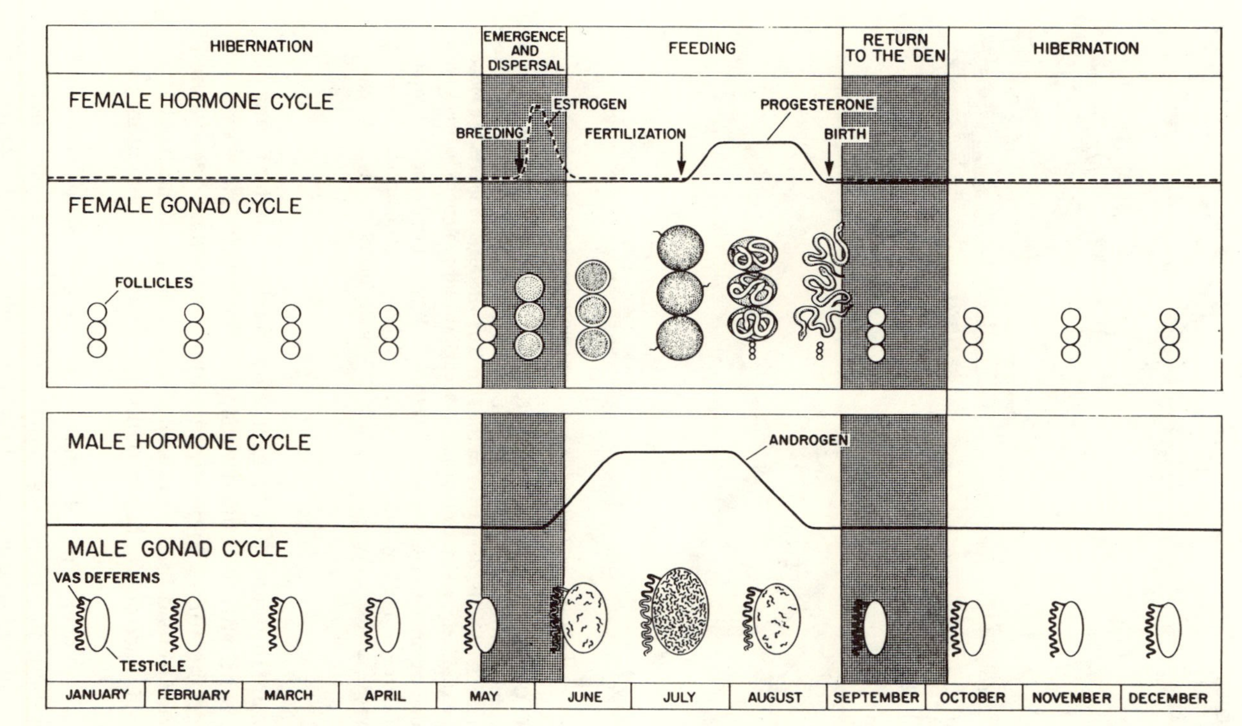

MISMATCH of physiology and behavior characterizes the reproductive behavior of the red-sided garter snake. From January through early May the snakes are in their den. In the female the blood level of the sex hormone estrogen is low, and the gonads (the ovaries) contain only small egg cells (follicles) lacking a yolk. In the male the blood level of the sex hormone androgen is low, and the gonads (the testicles) are small. The male's vas deferense, or sperm duct, is packed with stored sperm. The snakes emerge and mate late in May. Their gonads are still small and their sex hormones are still at an ebb. Only after mating are changes observed. In the female the mating causes the level of estrogen to rise. In response the eggs grow large and are filled with yolk. In the middle of July the eggs are fertilized by sperm the female has stored for six weeks. Then the level of progestorone, the pregnancy hormone, rises. In the male the level of androgen starts to rise at a time when the females have left the vicinity of the den. During the summer the testicles grow large and produce the sperm the male will need the following spring. In August or early September the female gives birth, and by about the end of September both the male and the female have returned to their den.

Figure 5.7b. A *Scientific American* graph with iconic illustrations. Redrawn from an illustration by Patricia V. Wynne from "The Ecological Physiology of a Garter Snake," by David Crews and William Garstka, *Scientific American*, November 1982, pp. 162–63.

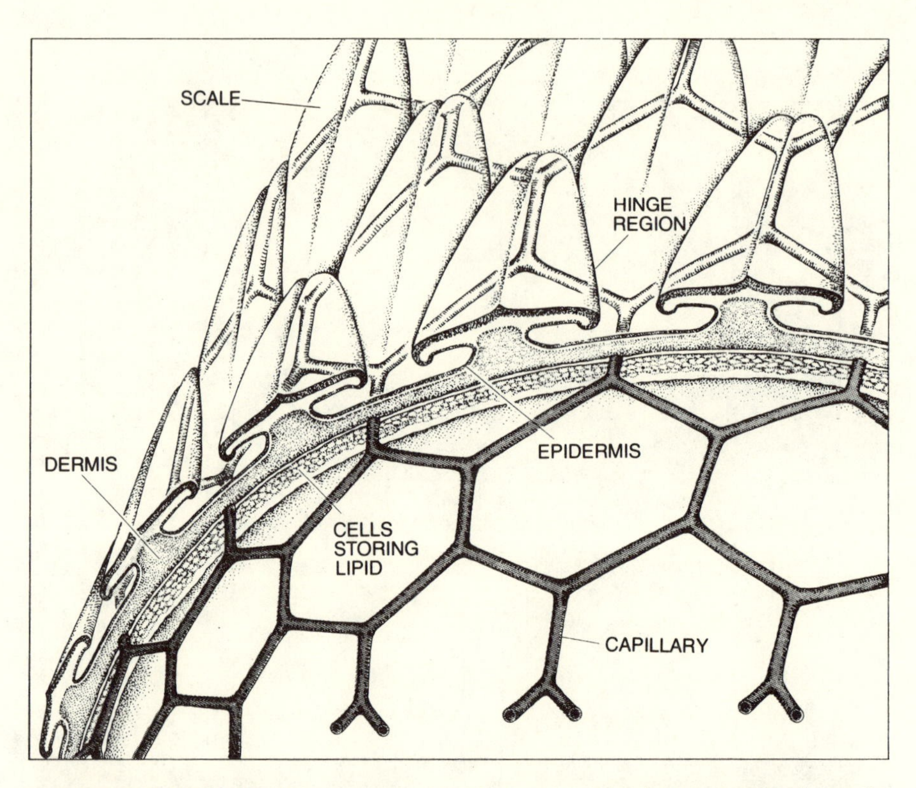

PATH OF VITELLOGENIN onto the back of the female so that it attracts males during the mating season is deduced from the presence of a network of capillaries and of cells that store fatty molecules in the dermis, the deep layer of the skin. The vitellogenin leaves the blood as it flows through the capillaries and then percolates through the hinge regions between scales.

Figure 5.8. A cutaway drawing. Illustration by Patricia V. Wynne from "The Ecological Physiology of a Garter Snake," by David Crews and William Garstka, *Scientific American,* November 1982, p. 168. Copyright © 1982 by Scientific American, Inc. All rights reserved.

That there are differences is suggested by an editor's letter to one of the authors, sent with the edited version of the article. He seems to have expected the author to be surprised by the revison, because he asks the author to read the letter before looking to see what has been done to his manuscript. He has two different lines of explanation:

> As you will see, we feel that it is necessary for the article to come to grips with the main subject somewhat more quickly. If that is not done, the reader may lose interest and the battle of popularization will be lost at the outset.

> We also feel that a fair amount of simple-minded explanation is necessary in order to make your argument fully accessible to the general reader.

The question such changes raise is what the main subject *is* that the article is supposed to come to grips with, and what part of the argument has been made accessible.

Organization

We saw the differences in structure in the comparison of popular and professional articles; in the revisions, we find that the authors tend to organize their manuscripts with simultaneous elements as in their articles for professional journals, whereas the editors tend to bring out the narratives focused on the organisms. The authors set their findings in the context of disciplinary history and concepts, whereas the editors emphasize the direct confrontation with nature.

The opening paragraphs of the three manuscripts all differ considerably from the published versions. Interestingly, since scientists are often accused of lacking historical context, both Gilbert and Parker begin their manuscripts with bits of scientific history. Gilbert, after a brief visual image of the butterfly in the jungle, spends two paragraphs describing how the work of H. W. Bates on these species helped to support Darwin's theory, and how recent work of Ehrlich and Raven developed the concept of coevolution. Parker also begins his manuscript with a historical review of sexual selection since Darwin. The authors may have been following the conventions of articles for professional journals, which usually include a brief review of the literature in their introductions, placing the current work in that context. For the *Scientific American* editor, such a context is a distraction. When the *Scientific American* version does give some of the development of these concepts, the passage is prefaced with an apologetic statement: "To answer the question, we must turn to a bit of history. . . . "

In both cases these introductions are cut, presumably, as the editor says of one of the articles, to get to the point quickly. The point, for the editor, is the organism itself: Gilbert's butterflies, Crews's snakes, and Parker's dung flies. Thus the editor of Gilbert's article revised his manuscript opening to mention the interactions of plants and animals in the first paragraph and the concept of coevolution in the second. Gilbert then revised this revised introduction to begin with the idea of coevolution, giving some well-known examples in the second paragraph, and finally coming to the *Passiflora* and the *Heliconius* only in the middle of the third paragraph. What is negotiated in the back-

and-forth of drafts is the proper way to introduce the research: *Scientific American* sees the concept of evolution as a way of explaining "one of the most remarkable interactions" of animals and plants, whereas Gilbert sees the concept as the "main subject," and the plants and animals as a way of studying it.

The cuts throughout Gilbert's article either move the focus away from evolutionary concepts and methodological issues, or sharpen the focus on the plants and the animals. For instance, the editor cuts Gilbert's statement of the null hypothesis in his outline of experimental design of leaf shape, his statement that understanding of the effect of *Heliconius* on *Passiflora* is only qualitative, his comments on the randomization of leaf mimics, and his hypothesis about the possible evolution of other antilarval traits in the species. All these passages introduce concepts and procedures of biology.

The narrative of the butterfly searching for a vine that I have used in the previous section as an example of the narrative of nature is almost entirely the creation of the editor, who reorganizes eight paragraphs of information into four, and sequences them around stages of searching numbered 1, 2, 3.

> To answer this question one must understand three aspects of the interaction between the butterfly and the vine. The first is how the female butterfly finds the host plant. The second is how the butterfly makes a choice between depositing its eggs or not depositing them. The third consists of the factors that affect the survival [of] the eggs and the caterpillars after they are in place on the vine.

The evolutionary point is explained in terms of this sequence of search and evasion: "Natural selection . . . would favor the mutant vine that was harder for the butterfly *to find,* that was less likely to be selected for egg-laying *once it was found,* and that was inhospitable to the butterfly's offspring *once they were hatched.*"

The correspondence between Crews and Garstka about their *Scientific American* article shows that the two authors discussed the implications of a temporal organization versus a thematic organization. Garstka, responding to Crews's first version, raises some "disagreements" about the emphasis on adaptation, and then goes on to raise questions about the organization:

> I strongly feel that the paper needs a central focus. The paper can't and shouldn't be built around a gee-whiz story of adaptation. I used post-nuptial gametogenesis in the previous draft, and here I've

> tried to follow a temporal progression, at least in the female stuff. . . . The paper should stick to the point, i.e. the animal, and not include other data (melanogaster and radix).

But their disagreement is not about whether the article needs a point, but about whether the animal is the point, or whether the point is rather a larger story of adaptation that includes these other species of garter snake, *melanogaster* and *radix*. Crews's outline of his preferred organization for Garstka stresses the progress of the research:

> The paper has four main section[s] in my view: description of the natural history and behavior, female sexual attractiveness pheromone, male inhibiting pheromone, and male sexual behavior—I am placing the male sexual behavior section after the perhormone [sic] stories because it is still incomplete.

The structure here is just the chronological narrative I have outlined, but the rationale for it follows his topics, not the mating process of the snake.

The negotiation between Crews and Garstka is like the negotiation between Gilbert and the *Scientific American* editor. Crews, like Gilbert and Parker, started with an introduction to the general concepts underlying his research: "Environmental extremes in temperature, food, and water require that animals have specialized physiological and behavioral adaptations to survive . . . " But Garstka revised the opening before the manuscript was submitted to start with the snake itself; in this he probably anticipated the preferences of the editor. Their general point is then in their fifth sentence: "The major problem we have addressed in our research is how the synchronization of physiology and behavior with environmental demands occurs in species that live in regions with extreme climates." The editor of *Scientific American* retains a version of this statement, but puts it in the twenty-first sentence, after the narrative of the hardships of the snake's life and the oddity of its mating behavior, which is expanded considerably from the manuscript. The extremes and images that draw in the general reader are emphasized by this rearrangement.

With the Crews and Garstka article, as in Gilbert's, the editor draws together scattered passages wherever possible to emphasize the narrative of the garter snake. For instance, on page 2 of the manuscript Crews and Garstka interrupt the story of the snakes mating to discuss the research of two herpetologists who showed that snakes will emerge only with warmer temperatures. The *Scientific American* version continues the story through the mating, the birth of new snakes, and the return to the den before discussing the herpetological research through which these events are known. As with Gilbert's article, the *Scientific American* editor has added adverbs suggesting time: "the males come out *first* . . . *then* the females emerge . . . all trying *simultaneously* . . . *meanwhile* the mated female embarks . . . *at the end* of the mating season. . . . *Early in the fall* . . . *then* the females and the males return."

The *Scientific American* editor also makes two small but striking additions to the Crews and Garstka article that are consistent with this organizational focus on the life cycle of the snake. The first paragraph of the *Scientific American* version ends saying "Their blood . . . becomes as thick as mayonnaise." Crews says this is based on something he mentioned in a phone conversation with the editor, but that the mayonnaise analogy was suggested by the editor. It is both strikingly effective and, Crews thinks, rather misleading; it has caused a number of questions and comments from other herpetologists. Later an added sentence says that when crows and ravens catch the exposed garter snakes they peck out the snakes' livers. This is rather distantly related to the topic of the reproductive physiology, and one of Crews's colleagues has commented in the margin of the *Scientific American* manuscript, "anecdotal." It doesn't fit with the research, but it does provide a striking detail for a narrative of the hard lives of these snakes.

Parker says in his cover letter to the editor that he tried, through many revisions, to make his draft of his *New Scientist* article "as interesting and conversational as possible." But he still follows, in his organization, the principle of his *Evolution* article, explaining how he deals with each term of a formula and each formula of his model, paying particular attention to how this apparently mathematical work relates to concepts in biology. The *New Scientist* editor makes changes that emphasize, rather, the researcher's own activity, on one hand, and the narrative of nature, on the other. For instance, Parker ends the introduction in the manuscript, "The approach I used was to make predictive models that could be compared with the observed strategies shown in nature by male dungflies." The introductory paragraphs of the *New Scientist* edited version end, "My approach was to make predictive models of optimal mating strategies. Then, notebook, pencil, and camera in hand, I set out among the cowpats to discover what the flies really did." The edited version is more vivid, but also changes the emphasis from a comparison of prediction and observation to simple observation of what the flies "really" do.

Most of the organizational changes by the *New Scientist* editor shift the emphasis away from mathematical concepts to simple observation. For instance, the first section of Parker's manuscript is headed, "Dungflies obey the input matching law," and the next is headed, "Models dependent on sperm competition." There are no headings in the *New Scientist* article. Most sections of the manuscript begin with a problem seen in terms of filling in the formula. "Many aspects of male dungfly sexual strategy are difficult to analyze unless we establish how the sperm from different males is used during fertilization." There follows a way of figuring mathematically what the chances are for each male. The editor replaces this with a simple question, "Why do males guard their females when they are laying eggs?" The effect of such changes, throughout the article, is to shift the emphasis. Consider the difference in meaning between these sentences from the manuscript and their revised version:

> With a suitable correction to include the effect of mating with virgins, this data can be used to construct a curve of expected gains (eggs fertilized) with time spent copulating (Figure 2). Gains follow an exponential law of diminishing returns with mating time.

The editor tries to avoid all this mathematics:

> But, as we determined in laboratory experiments, the benefits of prolonging copulation are subject to diminshing returns.

In avoiding the mathematics, the editor makes it seem that experiments reveal this relation without the need for mathematical interpretation.

Syntax

The editors' changes in the authors' syntax may seem to be merely matters of editorial taste, and often they are. But three kinds of significant changes are common in all three articles: (1) rephrasing of introductory statements as questions and answers; (2) rephrasing of compound and complex sentences into several more simple sentences; and (3) rephrasing passive and impersonal constructions in active voice. Each of these changes relates to the differing views of the authors and editors.

One of the most powerful syntactical patterns of popular science texts is the question and answer.[7] The reader is conducted through a

7. Kenneth Morrison analyzes this pattern in "Some Properties of Telling-Order Designs in Didactic Inquiry," *Philosophy of the Social Sciences* 11 (1981): 245–62.

sort of dialogue in which, ideally, each of his or her vague confusions is stated as a clear question, which then receives a clear answer. In such a format, the direction of research, the point of all these facts, is always clear. This direction is often clear to the researcher only in retrospect, as many episodes in the history of science tell us. The decision about which questions are to be asked in a discipline at any time is just what is at issue. The popular audience thinks of science in terms of large and universally agreed-on questions (How can we cure cancer? How can we build a satellite weapons system?). But these may have little relation to the way an individual research program proceeds. The effect of the question and answer structure of popular texts is to imply that in research, as in undergraduate education, the questions are always given, and that, as in undergraduate education, the answers must surely follow.[8]

Question and answer patterns are extremely rare in articles for professional journals; the question is usually implied by some lacuna in the literature as described in the introduction. For instance, Williams and Gilbert's *Science* article has no questions; the question is only implied when the introduction says, "it is difficult to demonstrate a causal connection between a plant characteristic and a particular selective agent." But Gilbert introduces the pattern twice in his manuscript of the *Scientific American* article.

> With respect to which traits of passion vines are mutant individuals likely to be more or less successful in avoiding attack by *Heliconius*?

> What traits of *Passiflora* have evolved to deal with *Heliconius* eggs and larvae that have appeared on the plant?

In introducing these patterns, Gilbert shows he is aware of the different needs of his popular readers. But the editor of *Scientific American* uses the structure ten times, to introduce almost every new line of inquiry. So, for instance, Gilbert uses this introductory statement:

8. The alternative view of scientific questions and answers is suggested by George Eliot's account of Lydgate's physiological research in *Middlemarch* (1870); he approaches a "more scientific" view of the body even though he makes the wrong assumptions and asks the wrong questions in his search for a primitive tissue. One famous account of disagreements within a field about appropriate questions that should direct research is James Watson's *The Double Helix*. The Norton edition of this best seller (ed. Gunther Stent, 1981) includes a number of essays and reviews commenting on the popularization of science.

> From a passion vine's perspective, the effect of discovery by *Heliconius* depends upon the plant's size and location.

The editor changes this to a question:

> What are the consequences for the passion flower of being parasitized by a *Heliconius* butterfly?

But this question arises only if one considers as relevant certain variations in the population of passion vines. The researcher is often primarily concerned, not with answering a question, but with making a question of what may not have been a question before.

Crews and Garstka's *Scientific American* article seems not to follow this pattern of introductory questions; they begin only two sections with a question in their manuscript, and the published version has only three such questions. But a closer look shows that many of their introductory sentences focus the inquiry in just the way a question would:

> *The question* addressed in our research . . . is how a species such as a red-sided garter snake, which lives in a region where the climate is extreme, comes to have its physiology and its behavior synchronized with the demands of its environment.

> Investigators . . . *have long been puzzled* by the fact that the skin of garter snakes is devoid of any obvious glands that might produce, store, or release such a chemical.

> *We did not know* how the vitellogenin gets from the blood to the surface of the skin.

> . . . the mismatch between the onset of mating behavior and the size of the ovaries *is paradoxical.* [emphasis added]

Crews and Garstka use key words suggesting a question or a problem to mark for the general reader the line their inquiry has taken, or seems to have taken, so as to weave together a number of seemingly scattered experiments into one string.

Parker uses the question and answer format consistently in his manuscript, before any editing. It includes six questions, one at the beginning of each new mathematical problem, so that the question comes to signal the beginning of a new topic. Even the conclusion, cut from the article, suggests a question: "How sexual selection might have operated on man was a problem that fascinated Darwin." As we have seen, the editor adds several questions at the beginning to catch the reader's interest. But, surprisingly, he does not keep Parker's apparently logical form, with a question for every section. This seems to be because the editor wants to emphasize the long narrative of the dung fly's mating, where Parker's questions focus attention on his mathematics. Parker breaks the narrative into more problems that the *New Scientist* editor, and perhaps the *New Scientist* reader, wants to deal with.

Another major syntactical revision in all three articles, the breaking of compound and complex sentences into more simple sentences, would seem to be a straightforward improvement of readability. Certainly the densely packed sentences of many scientific articles make them difficult to read. But often this dense packing is the equivalent, on the syntactic level, of the tendency to simultaneity we saw on the level of organization; the authors want to bring as many ideas as they can, at once, to support their assertion. If this is the purpose for such dense packing, then the simplification of syntax also alters the article's presentation of space and time.

At least eight times the editor rewrites a very dense sentence in Gilbert's *Scientific American* manuscript as two or even three sentences. For example, this sentence, though grammatically quite correct, requires the assimilation of background information and the organization of two categories of processes, within a contrast, within a cause/effect structure:

> Because most species placing eggs singly are cannibalistic as larvae, females adding eggs to both occupied and unoccupied shoots at random will leave less progeny than females possessing egg avoiding behaviors.

The editor unpacks this into an entire paragraph:

> To consider predation, the emerging caterpillars of most *Heliconius* species that deposit single eggs are cannibalistic. One may suppose, then, that a major criterion affecting the decision of a female of these species to deposit eggs or not to deposit eggs would be the presence of another female's eggs at the selected site. A mechanism favoring the avoidance of such sites could easily evolve because mutant butterflies with such a mechanism will have more

numerous progeny than butterflies that deposit eggs at occupied and unoccupied sites randomly.

The editor's version gives the background in one sentence, the comparison in another, and the cause and effect in another. I assume it is easier to read (though some nonbiologists reading it have disagreed with me). So why do scientists persist in writing more tightly packed sentences? Here Gilbert wants to do several things *at one time:* to limit his statement to certain species who place eggs singly, to take into account their cannibalistic tendencies, to compare two patterns of female behavior, involving two appearances of vines, and to put all this in an evolutionary context. The structural terms that join the clauses and phrases in this complex sentence—*because, and, less . . . than*—are the key terms in an argument like Gilbert's, linking all his isolated observations and findings into a general explanation. The *Scientific American* version subtly alters his meaning by putting the decision of the butterfly first and the evolution of this mechanism second. It contributes to the sense a reader gets in the article, and in most popular articles in this field, of the purposefulness of evolution for the individual.

In the Crews and Garstka article, as in the Gilbert article, the most common change by the editor is the breaking up of single sentences into two or three shorter sentences. Ten of the twenty sentences in *Scientific American's* opening narrative are under eleven words long; only three of the first twenty of the sentences in Crews's manuscript are this short. As with Gilbert's article, the breaking up of these sentences subtly alters the relation of observation to concept. For instance, this sentence is hard to read partly because a complex process, and the location of an organ, are explained in a subordinate clause and an adjective phrase in a statement about researchers:

> John Kubie and Mimi Halpern have shown that the tongue-flicking investigation of the female's body by the male delivers pheromone molecules to the male's vomeronasal or Jacobson's organs situated on the roof of the mouth.

The rewritten version separates, syntactically, the observation of nature from its conditional statement as part of research, and transforms nominalized behavior (*tongue-flicking investigation*) into an active verb:

> The work of John Kubie and N. Mimi Halpern of the Downstate Medical Center of the State University of New York suggests how

> the pheromone acts. The male catches pheromone molecules on his tongue, which he repeatedly flicks as he nears the female, and the tongue delivers the molecules to the vomeronasal organs, which are on the roof of the mouth. The chemical-sensitive cells of the vomeronasal organs send signals over nerve fibers to the brain.

Again, this is easier to read, but it takes a statement about nature out of the frame of the research that demonstrates it.

The third major category of revision of syntax, the transformation of nominals and passives into active constructions, has frequently been discussed in analyses of scientific prose. Both *Scientific American* and *New Scientist* remind their contributors, who have become accustomed to the passive constructions of scientific journals, to use the active voice wherever possible. Still, Gilbert writes a sentence in this form:

> When branches of the host plant having similar oviposition sites were placed in the area, no investigations were made by the *H. hewitsoni* females.

The *Scientific American* editor rewrites this as:

> I collected lengths of *P. pittieri* vines with newly developed shoots and placed them in the patch of vines that was being regularly revisited. The females did not, however, investigate the potential egg-laying sites I had supplied.

Some readers see the second, active voice version as more realistic because it emphasizes the intervention of the scientist. But it emphasizes only his activity, not the conceptual framework he brings. Also, the second version, in making two sentences to describe his action and the butterflies' response, makes the claim a narrative.

Gilbert does not simply slip into impersonal constructions from an article-writer's habit; we can see his preference for them in looking at his comments on the *Scientific American* editor's version. For instance, he uses a long and rather difficult impersonal construction as the subject here in the manuscript version:

> The observation that inexperienced females are strongly attracted to wire models of tendrils . . . suggests . . .

Scientific American revises this to require Gilbert to supply a personal subject:

> For example [WHAT INVESTIGATOR OF WHAT INSTITUTION?], working with *Heliconius* females in the laboratory, showed that they were strongly attracted to wire models of passion vine tendrils. This behavior suggests that . . .

Gilbert, in his revision of the revision, changes this back to an impersonal contruction:

> Studies of young, inexperienced *Heliconius* females, carried out by Peter Abrams in my laboratory, showed that . . .

The information here is the same, but the emphasis is insistently on the studies rather than on the investigator.

Crews and Garstka have active sentences more frequently in their manuscript. But in their case, the *Scientific American* editor makes revisions that seem to have just the opposite effect from those in Gilbert's article. Here the manuscript version attributes a finding to the researchers in another field:

> Molecular biologists have established that estrogen acts on the fat bodies to induce the mobilization of stored phospholipids . . .

Scientific American rewrites this with the natural substance itself as the subject:

> They [phospholipids] are released into the blood when estrogen acts on the fat bodies.

So the mere proportion of active sentences will not tell us the degree to which the article emphasizes the activity of the scientist; here the editor brings out the narrative of nature, not the narrative of science, by making the sentence passive.

In Parker's article, too, revisions of syntax alter the meaning of the statements. One addition of a personal element that makes a difference in how we read the article occurs where he makes a strong claim for the relevance of his findings to natural selection:

> There can be no doubt that the behavior of male dungflies, with its intense struggles between males for females, offers impressive qualitative evidence for Darwinian sexual selection.

The *New Scientist* version keeps the statement, but in a qualified form:

> It seemed to me that the behavior of male dungflies, with their intense competition for females, offered impressive evidence for sexual selection.

In a scientific article, such a qualification would be a serious weakening of the claim, but here it seems to be added just to keep the personal tone and the detective-like narrative. (Note also that the *behavior* has the struggles in Parker's version, not the animals, and that the *New Scientist* version deletes Darwin.) It is interesting that Parker actually changes one of the editor's passive sentences back to active, questioning whether this sentence needs to be in the passive: "A hind's interest in sexual encounters . . . terminates once she has been mated with just once." Such a description of mating reminds us how loaded the active/passive distinction can be.

Terms

The most frequent changes in revision by the editors, and the changes the authors notice most, are the substitutions of popular terms for scientific terms. The substitutions may not change the informational content of the article, but some of them may change the narrative, again shifting attention from the narrative of science to the narrative of nature. For instance, the *Scientific American* version of Gilbert's article changes *oviposition* to *egg-laying* throughout, to Gilbert's annoyance. Certainly more people know the words *egg* and *laying* than know the word *oviposition*, and certainly the substituted phrase means the same thing. But *oviposition* is one of the many technical terms that changes a process into a concept. It may confuse some people, but it allows scientists to talk about this action as a category of behavior, as an entity in itself. Consider the similar term that begins this sentence in Gilbert's manuscript:

> Germination, and therefore small plants, occur[s] in forest gaps where disturbances such as treefalls and landslips have exposed the soil to sunlight.

The *Scientific American* version of this sentence changes the noun to a verb:

> A passion vine seed can germinate only on open ground where the soil is exposed to sunlight.

The *Scientific American* editor avoids the awkward ambiguity about the number of Gilbert's compound subject, but does so by making the *seed* rather than the *germination* the subject. Again the plant is given a narrative of its own.

The problem such revisions raise for the biologist is that nominalization is what allows him or her to talk about processes rather than organisms. One cannot talk about evolution, for instance as something an organism *does*. There is a difference in meaning between the phrase "divergence in the visual appearance of sympatric vine species" and the more readable *Scientific American* version, "Where different passion vine species coexist they differ from one another in leaf shape." The more familar words, *different* and *coexist* do not have the evolutionary meaning of *divergence* and *sympatric;* they do not suggest how these relations in the population came to be as they are through variation and selection.

Gilbert notices the tendency of the editor's changes of words to eliminate the activity of the scientist while foregrounding the direct reading of nature. The editor's version starts a new paragraph, after making the evolutionary point that mutant individual *Passiflora* having features that resist the *Heliconius* will contribute a larger proportion of seedlings to the next generation, with the transitional phrase, "This being self-evident . . . " Gilbert changes the phrase on the editor's typescript to "This being the case . . . ", and comments in the margin, "It sure isn't self-evident until you make the observations." Gilbert makes a similar change when he deletes the editor's *of course* in the sentence, "This suggests, of course, that the pressure of *Heliconius* parasitism has favored the evolution of passion vine leaves that deceive the female butterfly." Ironically the popular narratives, which often try to build up the authority of the scientist as a genius with an immediate relationship to nature, often end up leaving the genius with little work to do.

The terms in the Crews and Garstka manuscript are also changed in the editor's revision. Most of the technical terms—pheromone, cloaca, vesicles—are simply defined in appositives or in parentheses. For the more detailed accounts of experiments on hormones, many definitions are needed. These definitions show how much scientific texts depend on certain terms; the reader cannot follow the narrative of the experiments unless he or she knows, not just what these substances are, but why they would figure in this experiment, why they might be thought by other researchers to initiate sexual behavior in male garter snakes. Only a few of the changes of Crews and Garstka's words make the kind of conceptual difference that those in Gilbert's article do. For instance, the phrase "preceding testis growth cycle" is changed to "when the gonads are growing the previous year." This substitutes a familiar concept (*year*) for an unfamiliar one (*growth cycle*), but misses the concept of a cycle, the concept Crews illustrates with his graphs.

The *New Scientist* editor also finds that he needs to substitute more popular expressions for some of Parker's scientific terms: *gravid* becomes *heavy with eggs*. Some of the changes of vocabulary alter the reader's view of evolution. The manuscript has the definition of *sexual selection* in this sentence:

> Any inheritable characteristic that increases a male's mating chances should increase in the population because the male will have more offspring than other males.

The editor, in trying to simplify this statement, removes the element of competing for mates, leaving simply natural selection:

> any inherited characteristic that helps an individual produce more offspring than its competitors will become common in the population.

Similarly, Parker, in reviewing this revised version, changes the editor's phrase in the last sentence, which says that "behavior patterns evolve through passive selection of the most successful strategies." The word *successful* suggests a "survival of the fittest" tautology; Parker changes the phrase to say instead that evolution favors the most "fruitful" strategies, those which produce the most offspring. In another of the editor's revisions, "What determines the spatial distribution of searching males?" becomes "The males also face the problem of where to mate." The scientific concept is turned into the organism's narrative.

One crucial term that is all but eliminated is the *Evolutionarily Stable Strategy*, a concept at the heart of Parker's mathematical model. Parker introduces the term with a definition and an attribution: "To use a term developed by John Maynard Smith of Sussex University, we need to find an 'Evolutionarily Stable Strategy' (ESS)—i.e. a strategy which, once established, is unbeatable in evolutionary terms." *New Scientist* cuts the term, and the name of Maynard Smith, the originator of the mathematical approach to evolution furthered by Parker. Instead the editor substitutes, where necessary, the phrase "optimal strategy." That seems to be close in meaning, but again, the ESS specifies a process taking place in an entire population over the course of evolution; optimal implies the choice of an individual over a lifetime. The article uses the author's phrase only once, referring to the "so-called evolutionarily stable strategy." This is one of the devices for making the jargon seem unnecessary. Where possible, Parker changes "optimal" back to "evolutionarily stable" in reviewing the editor's typescript.[9]

Forms of Popular Science

Further study would be needed to show whether the distinction between the narrative of science and the narrative of nature applies to popularizations written by science journalists in general-interest magazines and newspapers. After all, if the narrative of nature were found only in such relatively sophisticated publications as *Scientific American* and *New Scientist,* it would not necessarily be very important in describing the views of the public as a whole. The articles in *Scientific American* and even in *New Scientist* are closer in form to scientific articles than are the reports in the science sections of the *New York Times* or the *Guardian* (which have excellent science reporting), or in *Time* or *Newsweek* (which are more sensational), or in the general news sections of newspapers. Articles in these publications are by science journalists, or journalists with no scientific background; they must compete for the readers' attention with daily headlines, football scores, and pictures of celebrities. As a first step toward applying the two narratives more broadly, as poles to which other articles can be related, let us consider more popular reports of the findings of Crews and Gilbert.

In chapter 4 I discussed the controversy following the publication of Crews and Fitzgerald's article in *The Proceedings of the National Academy of Sciences* titled " 'Sexual' behavior in parthenogenetic lizards (*Cnemidophorus*)." I have noted that both Cole and Crews published *Scientific American* articles on *Cnemidophorus.* But there were other popular texts responding to the news after the *PNAS* article. For instance, *Time* reported Crews's research in an article titled, "Leapin' Lizards! Lesbian reptiles act like males." The titles show how different the articles will be; the *PNAS* article puts "sexual" in cautious quotation marks, and gives its scientific name, while the *Time* article plays up parallels to human behavior. The *PNAS* article, like the other scientific articles studied, begins by presenting the findings in the context of the concepts of biology: "All-female, parthenogenetic species pres-

9. I have developed the syntactic and lexical comparison between scientific and popular articles in more detail in an unpublished article, "Lexical Cohesion and Specialized Knowledge" *Discourse Processes* (forthoming).

ent a unique opportunity to test hypotheses regarding the nature and evolution of sexuality." The *Time* article begins, "Readers of science journals know a good deal about bisexual aphids, 'homosexual' gulls, and 'transvestite' fish." The emphasis is on the strangeness of nature, but also on its accessibility through familiar concepts, even through familiar human stereotypes.

The subjects of the sentences in the *PNAS* article, like those in the *Science* and *Evolution* articles used for comparison, tend to be nominalized activities: "The initial observation. . . . Dissection revealed that. . . . A female-female mounting was also observed." The article does not strictly follow the Introduction—Methods—Results—Discussion format; methods are relegated to a caption, and a strong sense of chronology is retained. But it reads coherently because it focuses on the narrative of the scientist (the level of the narrative of the study in chapter 4), from collection, to observation of the activity, to dissection, to observation of related species. *Time* has two kinds of narrative, an opening and closing in which the sentence subjects are always the researchers or the readers of research, and a central section in which the subjects are the lizards. The central section is quite scientific in tone:

> An active female mounts a passive one, curves the tail under the other's body, strokes the partner's back and neck, and rides on top for one to five minutes. The active female lizard always has small undeveloped eggs, while the passive female has large pre-ovulatory eggs. But there are cyclic variations in behavior and egg size in these reptiles, and the roles reverse.

But there is no discussion of the experimental arguments supporting the observation of such behavior. The framing narrative is not the activity of the scientist, but the comedy of a scientist humorlessly watching lizards mate. Crews's own words are satirized:

> Says Crews, "We are now trying to determine whether this male-like behavior facilitates reproductive function." Translation: the psychobiologist does not yet know why the females mock the male-female behavior of related two-sex species.

Note that *Time*'s translation is not easier to understand than Crews's version; the translation is a way of characterizing scientific jargon as a smokescreen. The paragraph on theoretical implications stresses the dubiousness of any theorizing: "It is too early to announce. . . ." Real science, in this view, is just watching and waiting. The *Time* article reassures its readers that Crews, though he makes startling discoveries, doesn't really know anything they don't know. Crews is special, rather, because he has immediate contact with nature. He is an explorer to be admired for his feats.

The narratives of other news articles provide further evidence for Mayr's assertion that science journalism is more interested in discoveries than in concepts.[10] One example is the profile on Crews recently published in an *Esquire* issue devoted to "The Best of the New Generation: Men and Women Under Forty Who Are Changing America." The entry presents his work as that of discovery and practical application:

> Crews' discoveries about the reproductive patterns of reptiles have challenged some time-honored assumptions about human sexual behavior. Based on his studies of the all-female whiptail lizard, Crews theorizes that "sexual behavior" preceded the evolution of sex. He has established that certain male behavior can both stimulate and inhibit female ovarian growth, a discovery that helps explain why overcrowded animals often experience a drop in birth rates. "Crews has unraveled an important piece of information with application to humans—that the human brain has the potential to go in either a male or female direction," says Dr. John Money, professor of medical psychiatry and pediatrics at Johns Hopkins University.

Though time-honored assumptions are mentioned, it is not suggested that Crews is trying to transform the concepts of his own field, but rather that he is working against unscientific superstition. The biological concepts, in the popular view, must have been scientifically established. He is put in two familiar roles: the theoretical iconoclast who challenges unscientific prejudices, and the practical medical technologist who produces work with applications to humans.

A more restrained example of such popularization in general-interest publications is an article on Gilbert's work in the (London) *Times* in August 1982. Pearce Wright, the Science Editor, presents only a summary of the *Scientific American* article, but he selects from it and rearranges to make it appeal to a casual reader. As in the *Time* article, the organisms studied are anthropomorphized; the headline reads, "Deceiving vine keeps butterflies at bay." Gilbert's work is mentioned briefly, emphasizing his persistence ("ten years of field

10. Mayr, *Growth of Biological Thought,* p. 23.

observations and controlled experiments in botanical conservatories . . . "). But the article's angle is to present Gilbert as a Darwinian with a "theory."

> At a time when controversy is rife about the theory of man's evolution, Professor Gilbert, director of the Brackenridge Field Laboratory, University of Austin, and a postgraduate of Oxford University, indicates that the idea that plants and animals interact on each other to influence their respective path of evolution may raise eyebrows.

Controversy is not rife about the evolution of man among population biologists, but it is rife in Gilbert's home state; the *Times* presents Gilbert as he would be seen by the public, not as he would be seen in his discipline. As in the articles on Crews, the disciplinary and conceptual basis of the work is excluded.

A recent article on Gilbert in the University of Texas alumni magazine *Alcalde* provides another example of this exclusion. The cover has a lovely color photograph of the *Heliconius*, and the article by Don Massa focuses on the organisms Gilbert studies, on the fact that he works in greenhouses, and on the possible economic applications of his work. Evolution is mentioned only in passing. Ironically, the article ends with a quotation from Gilbert that neatly defines the journalist's own approach: "If you work on butterflies as model organisms, people have difficulty seeing past how you study to ask what scientific questions motivate the work." Gilbert implies that a researcher will have special problems popularizing the scientific issues when the model organism is fascinating and beautiful. But as we have seen, the same sort of focus is possible in articles on such neglected and potentially unappealing animals as lizards and dung flies.

Gilbert's comment about misconceptions of his particular sort of research raises the question of whether we could expect to find the narrative of nature in popularizations of other scientific fields. Some biologists complain that popular biology articles are more heavily edited than those on physics in the same journals, that the physicists get to keep more of their vocabulary and physical constructs, partly because there are no familiar alternatives available. It may be that research on subatomic particles is harder to fit into a narrative of nature than research on butterflies and plants.[11] But there are other devices for

11. Gilbert makes this point in his response to a passage in my manuscript. "The terms of physics have no connection to familiar objects. Editors have no handy alternative to 'quark.' The same can be said of molecular biology. What do you do with

making the objects of physics seem immediately perceptible and unambiguous, such as pictures of the equipment, emphasis on the technology, or shaded diagrams of spherical particles. What I have called the narrative of nature draws on a tradition of natural history writing, but there are several other traditions of popularizations, such as the dialogue, or the detective story, that have similar effects, in foregrounding the activity of the object and obscuring the activity of the scientist.

Understanding the Public Understanding of Science

I have shown that the popularization of texts in evolutionary biology does not involve merely translating some technical terms, substituting active for passive voice, and focusing on some angle of popular interest, but in effect turns one sort of narrative into another. The narrative of science and the narrative of nature remain consistent, and consistently different from each other, because they support two different views of science. As I noted in chapter 1, a number of studies have shown how the form of the scientific article embodies the assumptions of the scientific community about the impersonality, cumulativeness, and empiricism of scientific knowledge. Such texts function to integrate researchers and their findings into the work of the research community. Researchers show that their findings are real because they meet disciplinary standards for methodology, they fit their work within disciplinary concepts, they submit the personal point of view to certain constraints. Each article is a demonstration of the need for scientific expertise.

The popular texts support an equally coherent and definite view of scientific practices, but one that is inconsistent with the view embodied by the scientists in their articles. In this view the scientist is alone, and proceeds without concepts or methodology, by simple observation of nature. There are no choices to make about the course of research, which proceeds from given questions to unambiguous answers. Just as scientists have an interest in promoting scientific expertise, the public, and those who edit journals with the public in mind, have an interest in this view, which minimizes expertise and emphasizes the unmediated encounter with nature. All scientific knowledge is brought within the

---

'restriction enzyme' or 'transposon'? I objected to having 'petiole' (a word found in Webster's) changed to 'leaf stem.' Maybe it gets the idea across but I didn't care to be laughed at by my colleagues!"

Gilbert and Mulkay, in *Opening Pandora's Box,* p. 168, show a textbook picture in which biochemical processes have been anthropomorphized.

realm of common sense, all scientific knowledge serves public goals. Articles for the public, when they are well edited to appeal to a general audience, reproduce science as the public wants it to be.

The effect of popular articles on the public perception of science is apparent in public discussions of scientific authority. The century-old public controversy over Darwin that the *Times* editor mentions is perhaps a reflection of the tension between these two views. Evolution has been presented as a fact of nature, so the public thinks it must be unchanging, but it is also a concept of science, so it must be open to modification. When one of these biologists calls evolution a theory, he means it is a central disciplinary concept enabling further thinking about life. When the *Times* calls it a theory, the connotation is that it is another airy idea dreamed up by scientists; there is no place for theory in popularizations. In Texas, Gilbert's and Crews's home state, the state textbook commission for a time banned the use of the word *evolution* in biology textbooks except when it was labeled as a theory. I would argue that, in a more subtle way, popular science texts, and even those on Darwinian topics, tend to exclude evolution because of the way they tell their story. As long as the popularizations focus on individual organisms, a concept like evolution is very difficult to imagine.

The effect on scientific discourse of the split between these two narratives is less easily traced, but it may be equally important. My study tends to present the popular articles as versions, more or less accurate, of the professional articles. But D. R. Crocker, a biologist comparing best-selling books on ethology with academic books by the same authors, takes the interesting view that the popular texts have priority:

> I suspect the authors let their genuine feelings spill out into their nature books and that academic pressure to be objective simultaneously encourages them to dissemble. My bet is that the popular informs the academic rather than the other way around.[12]

As I suggested at the beginning of the chapter, I would not give the popular view priority in this way. The complex form of professional scientific texts cannot be explained merely as the result of academic pressure to be objective. But the ways in which the popular could influence the professional versions may be suggested by an anecdote

12. D. R. Crocker, "Anthropomorphism: Bad Practice, Honest Prejudice?" *New Scientist* 16 July (1981): 159–62.

told by one of my subjects. He heard that a senator was about to make his research—which can be made to sound pretty silly—an example in an attack on wasteful spending; he was told by sources who knew the senator that he had to explain in a letter of one page (the senator would read no more) just what his work had to do with the larger good of the nation. This letter might be taken as an unusually direct example of the public relations function of popular science. But this and other popularizations are not just exercises in persuasion; to find such words is also to find a place in the larger society outside science, and in some degree to adapt the research to it. Both Terry Shinn today and Ludwik Fleck fifty years ago have stressed the important part that writing for a general audience plays in the production of scientific knowledge.[13]

To show how the categories "narrative of nature" and "narrative of science" might apply to other texts, let me take my own paper as an example. I could rewrite it as a fairly straight sociological paper in an Introduction-Methods-Results-Discussion format, setting out my findings against the background of other sociological studies of popularization. Instead, I find as I rewrite it that it is becoming more and more like a popular article. Note, for instance, that I have focused on the scientists, just as they focused on the butterfly and the vine, or the snake, or the dung fly. I have arranged the texts in a sort of chronology, from professional article to manuscript of popular article to published popular article, giving the professional version the appearance of priority, in the same way the editors arranged the articles in the chronology of the animals involved. I have added introductory questions, and omitted quite a bit of sociological jargon. I have removed to the notes almost all my references to the work of other sociological discourse analysts.

The irony, then, is that the narrative I use works against the concept I want to present: it conceals the evidence of the construction of my own text within disciplinary practices. But I would prefer not to revise away this irony. Instead, I will use it to point out the same tension in the study of discourse as in the study of evolutionary biology: between assuring the world at large that there is an external object, totally accessible (in this case, the process of popularization), and assuring that the analysis of this object is possible only in the terms laid down by a discipline, only by someone with the proper expertise. In the cases of

13. M. Cloître and T. Shinn, "Expository Practice: The Social, Cognitive, and Epistemological Linkage," *Expository Science: Forms and Functions of Popularisation* (Dordrecht: D. Reidel, 1985), pp. 31–60; Ludwik Fleck, *The Genesis and Development of a Scientific Fact* (rpt., Chicago: University of Chicago Press, 1981), p. 118.

the articles by Gilbert, Crews and Garstka, and Parker, and in the case of my own study, the tension between these kinds of authority in popularizations does not become an issue for the reader because there is no controversy in the public forum that would prompt nonscientists to question the assumptions of the various disciplines, or to compare various claims to expertise. Such evaluations enter into popularizations in public controversies like those over the safety of nuclear power, the efficacy of star wars, the alleged relation between race and I.Q., or the claims of sociobiology—the subject of chapter 6.

Chapter Six

Narrative and Interpretation in the Sociobiology *Controversy*

When E. O. Wilson's *Sociobiology: The New Synthesis* was published in 1975, it was given wide publicity, in newspapers and general-interest magazines, as well as in the popular science press and the journals in biology and the social sciences. Most reviews by behavioral biologists welcomed it, viewing it as a classic survey of the biological literature that drew attention to important and rapidly developing work that had not been brought together before because it fell between several established specialties. But there were also immediate criticisms by researchers in comparative psychology, genetics, and anthropology, and by groups of scientists in the United States and in Britain who were concerned about the political implications of the book. The arguments have continued since then. The debate over *Sociobiology* offers the chance to study a controversy like that among *Cnemidophorus* researchers in chapter 4, but a controversy that goes beyond one group of researchers and enters the public forum. It also offers the chance to study the ways popularizations use narrative to make claims for the relevance and authority of the methods of various disciplines, and the ways popularizations can dismiss rival disciplines. Wilson's book is unusual in its size, in its format, in its place in the author's career, and in its audience—what other recent scientific book is both a synthesis for specialists and a polemic addressed to a wide academic audience? But the dynamics of the controversy are not unusual (we could compare it to many other controversies in the public realm), and it can help us understand how the discourses of various biological disciplines relate to each other, and how the discourse of biology relates to political discourse.

The literature on this book over the past ten years is rather daunting; I am drawing on half a dozen books and collections of papers devoted to it and about twenty or so reviews in journals, and I am aware that I am overlooking a great deal. There have been a number of methodological studies by philosophers, some good historical and sociological studies, and even some studies of studies.[1] Rather than rehash the arguments presented in these texts, I would like to look at the texts themselves. I am going to argue that part of the power of Wilson's book is in its narrative structure, and that the controversy that followed can be seen in terms of the interpretation and ironic reinterpretation of that narrative and other competing narratives.

What I've missed so far, in the mountain of reviews by supporters and critics of sociobiology, is an explanation of just why Wilson's book is so persuasive, why I, a nonbiologist whose ideas about society are quite different from Wilson's, and who read the book after years of reading only its critics, could read it from cover to cover, suspending for the duration my disbelief. To attempt an explanation of this rhetorical power, I am going to argue that Wilson in *Sociobiology* incorporates and transforms the conventional narrative of natural history texts, with their sense of an immediate encounter with nature, by stripping them of narrative elements and then reconstructing the fragments into a grand narrative of evolutionary adaptation. And I will treat the criticisms of the book in the same way, not as arguments that make their points more or less conclusively, but as texts that reconstruct *Sociobiology*, putting it in a new context, transforming its narratives, and accounting for it. I am going to use the texts sociologically to examine one instance of the processes through which the authority of science is established and is applied in public controversy.

When I say *Sociobiology* is a narrative I do not mean to imply that Wilson is doing something unscientific. (See chapter 4, notes 4 and 7 for a parallel problem.) I must emphasize this because both Wilson and his critics criticize the telling of stories as a resource in scientific rhetoric. Wilson's critics say he tells "Just-So Stories" of adaptation. And Wilson is at pains to separate himself from such popular sociobiological writers as Lionel Tiger and Robin Fox, whose works advocate hypotheses by "selecting and arranging . . . evidence in the

1. The most useful studies for my approach were by David Hull ("Scientific Bandwagon or Travelling Medicine Show?"), Gerald Holton ("The New Synthesis?"), and Ullica Segerstrale ("Colleagues in Conflict"). Segerstrale's article is particularly helpful in giving details of the complicated context in the discipline. The philosophical studies by Ruse, Burian, and Dunbar were also helpful. As I note later, Joe Crocker has a good analysis of the political critique of sociobiology, and W. R. Albury has a good analysis of sociobiologists' responses to the political critique, though I think he fails to apply a similar analysis to the critics. In "Sociobiology and Ideology: The Unfinished Trajectory" Martin Barker makes some comments on his earlier analysis in *The New Racism* that are relevant to some of the issues I discuss later in the chapter and in my "Conclusion."

most persuasive manner possible" so that "verbal skill . . . becomes a significant factor" (p. 28).[2] The verbal skill of many of the participants in the controversy, their selecting and arranging, is precisely what interests me. What makes Wilson's book different from those of popular writers like Tiger, Fox, Desmond Morris, and Robert Ardrey—and it is different—is not his method of "postulational-deductive model building" or "strong inference" (the terms he uses to describe his method) but his way of making his model seem to correspond with perceived reality. He does this by inserting a narrative of natural history, which we associate with reality, within a narrative of evolution, which we associate with model-building.

The Narrative of Natural History

Wilson is ambivalent about the power of natural history. He seems to use the term in a favorable sense when he begins his last chapter by asking us to look at man in "the free spirit of natural history." But through most of the book he uses the phrase to describe disciplines like sociology, psycholinguistics, or studies of mammalian behavior that he thinks have not yet developed out of the messy adolescent phase of inquiry to become mature sciences. Natural history is the opposite pole from developed theory; he warns that "natural history is sometimes so diverting, to the point of making one forget the main thrust of the theory" (p. 32). And it is indeed diverting: even such critics as the anthropologist S. L. Washburn and the sociologist Bruce Eckland admit their fascination with the details of animal behavior collected in the book.

For my purposes, natural history is neither a stage of disciplinary history, mature or immature, nor the subject matter of animals and plants, but a kind of text. Natural history gives a written account of actions of particular animals at a particular place or time, recorded by particular observers, as in this passage from Darwin's *Journal* of the voyage of the Beagle:

> I took the boat and rowed some distance up this creek. It was very narrow, winding, and deep; on each side a wall thirty or forty feet high, formed by trees intertwined with creepers, gave to the canal a singularly gloomy appearance. I saw there a very extraordinary

2. All references in parentheses without further details are to pages in Wilson's *Sociobiology*. Other works cited by short title in the text or notes are listed in the Reference List, section 5, "Texts Discussed: Chapter 6."

> bird, called the Scissor-beak (*Rhynchops nigra*). It has short legs, web feet, extremely long-pointed wings, and is about the size of a tern. The beak is flattened laterally—that is in a plane at right angles to that of the spoonbill or the duck. It is as flat and elastic as an ivory paper-cutter, and the lower mandible, differently from every other bird, is an inch and a half longer than the upper.[3]

The passage suggests some of the textual signals we can use to define the natural history narratives in *Sociobiology.*

- The use of the past tense, the tense used in English for particular moments in the past.
- The presence of apparently gratuitous details of time and place.
- The treatment of animals as individual, sometimes anthropomorphized, characters.
- The attention to the observer's perspective and response, and especially to what seems remarkable or strange.

The bits of natural history in *Sociobiology* that have these features are nearly all quotations from other observers, so one might think they were irrelevant to Wilson's own methods. But Wilson is unusually generous with such quotations, letting them have their say, and they play an important part in the texture of his narrative. Earlier natural historians comparing and classifying the forms of animals brought back specimens and had them stuffed and collected in museums. The quotations are Wilson's way of bringing back specimens of behavior.

The Past Tense

In natural history, things happen. Such events are indicated in natural history texts such as the passage just quoted from Darwin's *Journal,* by the use of the past tense. In contrast, the present tense usually indicates, in scientific texts, the general nature of the phenomenon being described, asserting that it is true at all times. The effect of the shift in tense can be seen in a quotation Wilson uses (p. 135):

> According to Schaller (1972), "Wildebeest sometimes stampede toward a river from as much as 1 km away. The long column of animals hits the river at a run, and if the embankment is steep and the water deep the lead animals are slowed down while those behind continue to press forward until the river turns into a lowing, churning mass of

3. Charles Darwin, *Journal . . . During the Voyage . . . of H.M.S. 'Beagle'* (London: John Murray, 1901), p. 146.

animals some of which are trampled and drowned. One such herd I observed at Seronera left seven dead behind; several hundred may drown in such circumstances."

The first sentence states a general fact, with an indefinite article ("*a* river") and a measurement of distance but not a location. The second sentence tells an exciting story, but in general form; we are to imagine this happening now and then, here and there, constituting one item in the behavioral repertoire of wildebeest. The first part of the third sentence shifts to past tense, to tell about *one* herd, which *I* observed, at a particular place ("Seronera") with a specific number of deaths as a result. Finally, after the semicolon, the event is generalized, again in present tense.

Such shifts to past tense narratives occur throughout *Sociobiology*, in accounts of birds mobbing (p. 47), gazelles stotting (p. 124), wasps fighting (p. 284), wild dogs adopting cubs (p. 125), or chimpanzees using tools (p. 173).

Use of leaves for body wiping. The Gombe stream chimpanzees commonly used leaves to wipe their body free of feces, blood, urine, semen, and various forms of sticky foreign material such as overripe bananas. "A 3-year old, dangling above a visiting scientist, Professor R. A. Hinde, wiped her foot vigorously with leaves after stamping on his hair" (Van Lawick-Goodall, 1968a).

That this is in the past tense marks it as a statement about one particular group of chimpanzees. When the general statement is followed by a quoted passage we would expect a particular incident supporting the general statement; the practical joke comes when we see it is also an embarrassing anecdote involving a particular (and eminent) victim, whose hair is apparently classed with overripe bananas as "sticky foreign material."

For a natural historian, even the Darwin of the Beagle *Journal*, the anecdotes are the point, and the scientific generalizations are framed within the narrative. Wilson, on the other hand, frames these narrative accounts in the present tense of scientific generalization. But the past tense particulars have their own authority, even in passages that seem quite theoretical. Compare the effect of these two sentences from the same paragraph, the first describing a mathematical model and the second describing an observation (p. 326):

> As Trivers has pointed out, there may come a time when the investments of both partners are so great that natural selection will favor desertion by either partner even if the investment of one is proportionately less.

> Rowley (1965) described a parallel episode in the Australian superb blue wren *Malorus cyaneus*. Two neighboring pairs happened to fledge their young simultaneously and could not tell them apart, so that all were fed indiscriminately as in a creche. One pair then deserted in order to start another brood. The remaining couple continued to care for all the young, even though they had been cheated.

The second passage describes something that was observed only once, but it is something that actually happened. Wilson does not offer it as any sort of proof, but it is persuasive nonetheless.

Gratuitous Details of Setting

Early natural history accounts were provided by travellers and explorers, so it is not surprising that a strong sense of place remains. For instance, in the Darwin passage, the fact that this bird was seen by a creek bordered with forest is ecologically relevant, but the references to a particular creek, to the height of the walls of vegetation, and to the gloomy appearance all go beyond a description of a habitat. Wilson gives gratuitous details of setting in many quoted and paraphrased passages; one example stands out because he uses it twice to show the scaling of dominance behavior (p. 444).

> When black iguanas (*Ctenosaura pectinata*) occur in less disturbed habitats, so that individuals are able to spread out, each solitary male defends a well-defined territory. Evans (1951) found a population in Mexico which was compressed into the rock wall of a cemetery. During the day the lizards went out into the adjacent cultivated fields to feed. At the rock wall retreat there was not enough space to permit multiple territories, even though the food supply in the fields was able to support a sizable population. As a result the males were organized into a two-layered dominance hierarchy. The leading male was truly a tyrant. He regularly patrolled his domain, opening his jaws to threaten any rival who hesitated to retreat into a crevice. Each subordinate possessed a small space which he defended against all but the tyrant.

The fields, the rock wall, Mexico, and the daytime are all relevant details of setting, but the fact that it is a *cemetery* wall is not ecologically significant. Still, it sticks in our minds, as it must have stuck in Wilson's, and surely it colors the highly anthropomorphic story that follows. Similarly, the lovely two-page drawings by Sarah Landry that illustrate Wilson's book often give a sense of a specific, named place, as well as providing general information about the conditions under which the animals live, what they eat, or what other species compete with them. Wilson's comments on frog calls give a powerful sense of setting in a style reminiscent of that of nineteenth-century naturalists: "The wailing of thousands of spadefoot toads (*Scaphiopus*) in a Florida roadside ditch, in the pitch-black darkness of a hot summer night, brings to mind the lower levels of the Inferno" (p. 443).

Animals as Characters

Animals described in biology are typical of their species, often distinguished only by the observable characteristics of sex and size. Animals in natural history are individuals like characters in novels, and they may even have names. Wilson notes that it is methodologically important that the primate ethologist can pick out individuals without marking them artificially, as must be done with the insects that Wilson studies. With chimpanzees and gorillas, "It is easy for observers to recognize individuals at a glance and even to guess their parentage with a high degree of accurancy" (p. 517). Gorillas are also recognized as individuals by other gorillas, and in one passage this recognition seems indistinguishable from that of the observer (p. 538):

> Fossey has stressed the importance of the personal idiosyncrasies of the dominant males, who control the movements of the group. One of the groups was under the control of Whinny, a silver-backed male given his name because of his inability to vocalize properly. When Whinny died, the leadership passed to the group's second silverback, Uncle Bert, who clamped down on the group's activities "like a gouty headmaster." Where the group had previously accepted Fossey's presence calmly, under Uncle Bert's command they changed to breast beating, whacking at foliage, hiding, and other signs of alarm. Soon they retreated into a more remote area higher up Mount Visoke.

The gorillas do recognize as different those individuals who have silver backs, but it is the observer who gives the names and the characterizations. The two kinds of individualization merge. Several other accounts of animals give names (pp. 214, 512), including that of the macaque Imo (p. 170), who is famous as the inventor of potato washing (and who is especially famous after she was discussed on David Attenborough's popular BBC series *Life on Earth*). She even gets an entry in Wilson's index, though she might be offended to find herself identified there as a chimp.

As the paraphrase of the Fossey account shows, anthropomorphization is common in natural history accounts. When it occurs in the eighteenth- or nineteenth-century texts that Wilson collects (pp. 281, 370, 542), one might imagine it to be an antiquated device. A particularly lovely passage by Guthrie-Smith on the sham death throes of pied stilts (one which is quoted admiringly by one of the reviewers) was written in 1925 (p. 123). But anthropomorphism also occurs in a number of the most dramatic passages from contemporary scientific texts that Wilson quotes (pp. 214, 473), as well as in his accounts of his own studies of ants, and in his coinages of new terms (p. 413).[4]

In discussing popularization, I noted D. R. Crocker's argument that anthropomorphization is not a bias added in popular texts to make them interesting, but is an unavoidable part of the scientific work of ethology that is more or less successfully concealed in the more scientific publications. One can see the human shaping of the animals' narrative in a passage Wilson quotes from an academic work by Alison Jolly (p. 278):

> "On August 16 and August 24, 1963, and in a more leisurely fashion, on March 23, 1964, a whole troop of *L. catta* barred the *Propithecus'* way, while the *Propithecus* returned their teasing. Again, the animals leaped towards each other, stared, feinted approach, but never came into contact. All the game lay in leaps and counterleaps, the *Propithecus* trying to pass through the *L. catta* troop, the *L. catta* attempting to keep in front of them, facing the other direction. Since there are about twenty *L. catta* to five *Propithecus*, the *L. catta* had an advantage; if one animal does not outguess the *Propithecus'* next move, another can do so."

Jolly's attribution to the lemurs of leisureliness, teasing, game-playing, and guessing remind us that human observers define behavior in human terms; it is the human analogy that enables us to see a series of actions as a behavior. The analogy is an old one; Wilson, with

4. Wilson responds to criticism of his anthropomorphic terminology in his *BioScience* article, "Academic Vigilantism."

his encyclopedic knowledge of eighteenth- and nineteenth-century writings on insects, reminds us of the tradition of moral fables around the ants and the bees. Ethel Tobach's highly critical review of *Sociobiology* says that Wilson concludes that "it is the social insects who have evolved the most peaceful and perfect of social organizations." And she comments, "The adage is old and worn." Perhaps it is only another way of observing the same thing to say that Wilson, in comparing the insects to humans, is not only drawing on the scientific tradition of his teacher, the entomologist William Wheeler, but is drawing on a much older literary tradition.[5]

The Perspective of the Observer

If animals are made into characters in the natural history narratives, so are their human observers, the travellers and autobiographers: Darwin rowing up the creek and being affected by the gloom of the forest wall, or the tiny figure of a painter that Thomas Cole paints in the lower left-hand corner of his huge canvas of the Oxbow in the Connecticut River. The reader is aware of observers in many of the natural history passages Wilson quotes; for instance, in the passage from Fossey that I have already quoted, the focus shifts from the observer watching the gorillas to the gorilla watching the observer. In another quoted passage, Alison Jolly recreates the observer's gradual construction of a scene (p. 530):

> "Your first impression of an *L. catta* troop is a series of tails dangling straight down among the branches like enormous fuzzy striped caterpillars. Later, with difficulty, you put together the patches of light and shade into a set of curved gray backs, of black and white spotted faces, of amber eyes. By this time, if the troop does not know you, they are already clicking to each other, and first one and then a chorus begin to mob you with high, outraged barks. The troop is quite willing to click and bark for an hour at a time in the yapping soprano of twenty ill-bred little terriers."

In the illustration of this scene drawn by Sarah Landry, the observer is made strikingly present by the gaze of the largest lemur; the caption says that "one male faces the observer with a threat stare" (p. 532). In

5. Tobach, "Multiple Review" (1976). One literary example of this tradition of moral or political fables is the charming story of the bees and the kingfisher that Hector St. Jean Crevecoeur presents as if it were a natural history observation in *Letters from an American Farmer* (1783; rpt. New York: Dutton).

another of Sarah Landry's drawings, showing a pride of lions, the caption tells us that one of the animals "stares at an unidentified object past the observer" (p. 506). If this were a photograph the object outside the frame would be unidentified, but here in a reconstruction it is, of course, imaginary. The caption, by treating the unseen object as real, emphasizes how the delimitation of the frame in these drawings makes them stand as representative of a larger world around them. In making this connection between particular observations and general statements about the world, the drawings and the natural history texts on which they are based follow the metonymic strategy some literary critics attribute to the realistic novel.[6]

There are a number of other textual means of making the reader aware of the observer, besides the pictorial devices of perspective and framing. The passage quoted from Jolly gives a sense of the observer by describing the development of her perception in time, as a narrative. Wilson admires other observers, who might be unseen in their own scientific reports of their observations, for the strenuousness and persistence of their work (p. 31), or for their exposure to danger (p. 495). Some of Wilson's illustrations present symbolically the problems of observation; for instance, one figure includes a human head to represent the observer of animal communication, and another includes two human heads, with angles of vision drawn from their eyes, to show the consequences of two observers having different definitions of populations.

We also become aware of the observer when he or she states a response, especially when they respond to the scene as something strange or extraordinary. Whereas biology texts focus on the representative and make all creatures ordinary by finding a place for them in biological description, natural history texts—like their BBC descendants today—focus on the remarkable or impressive. When Darwin comments in the passage I've quoted on "a very extraordinary bird," we are seeing Darwin as well as the bird. The natural history strain in Wilson's book is apparent in the striking pictures that show frogs, ants, or birds that are strange in appearance or behavior, and in his frequent expressions of awe at the wonders of nature:

6. I have analyzed the illustrations in *Sociobiology* in more detail in another paper, "Every Picture Tells a Story." For an influential analysis of metonymy and other rhetorical figures in the discourse of history, see Hayden White, *Metahistory* and (for a very concentrated presentation) "The Fictions of Factual Representation" in *Tropics of Discourse*.

> The males belonging to species on this list [of birds that mate after communal displays] are among the most colorful of the bird world. The brilliant red cock of the rock, for example, is easily the most spectacular cotingid, and the birds of paradise are justly considered the most beautiful of all birds. (P. 332, see also pp. 46, 220, 331, 332, 375, 423, 529)

The explanation of such colors by reference to sexual selection is characteristic of what I am calling biological texts; the response to one of these birds as beautiful is characteristic of what I am calling natural history texts. Though the bits of natural history are not representative of Wilson's usual style, and the natural history passages quoted make up only a small part of the text, they are the basis of its authority with the popular reader, because they connect all the model building to the immediate experience of nature. The quoted bits of natural history are like once-scattered specimens of behavior, all brought under one textual roof, not in the form of the emphatically unreal stuffed animals of museum dioramas, but in the form of stories.

Arrays of Information

If *Sociobiology* were just a massive anthology of natural history, it would not have aroused controversy. No observation, however awesome, horrific, or bizarre, is controversial outside some theoretical context. What makes *Sociobiology* dangerous or promising (depending on one's view) is that, like the museums Louis Agassiz or Richard Owen envisaged in the mid-nineteenth century, it projects a vision of the world, "an epitome of creation," as Agassiz's biographer called it. It seems appropriate to quote Agassiz's plan, since the Museum of Comparative Zoology that he founded now employs Wilson (as well as two of Wilson's most prominent critics).

> The casual observer . . . should walk through exhibition rooms not simply crowded with objects to delight and interest him, but so arranged that the selection of every specimen should have reference to its part and place in nature; while the whole should be so combined as to explain, so far as known, the faunal and systematic relations of animals in the actual world, and in the geological forma-

> tions; or, in other words, their succession in time, and their distribution in space.[7]

What makes Wilson's book different from those of popular sociobiologists, and also different from Richard Dawkins' presentation of sociobiological ideas in *The Selfish Gene,* is the way it builds up an immense array of representations of life like the halls of a museum. This is not just to say that *Sociobiology* is a big book, bigger than those of critics or popularizers of sociobiology, though bigness is part of it. Wilson does not merely collect a large number of the narratives I have described; he arranges them so that they keep their narrative force, their immediacy, while they are stripped of all their particularity, their excess details of sequence, time, place, and perspective; he transforms narratives into information. The narrative of nature becomes the narrative of science only by passing though this stage of stripping and arranging.

As I have noted, natural history texts seek out the singular, whereas biology texts seek out the typical. Wilson is careful to emphasize that a single observation, however careful, means nothing until it can be combined with others. For instance, he criticizes one ethologist, saying "Idiosyncratic actions of individuals do not constitute roles; only regularly repeated categories fulfill the criterion" (p. 299). And he comments at one point that "one anecdote does not prove the existence of a behavior" (p. 46)—even though, further down the same page, he makes skillful use of such an anecdote about parental care in monkeys.[8]

The text can make anecdotes into behavior by combining many observations through comparisons, classifications, or models. When Wilson compares three stages of "aggressive displays" in a figure of a monkey and a bird (figure 8-3, p. 180), he must leave out the developmental and behavioral narratives implied in telling when and how these two very different species make these displays. As the compara-

7. Elizabeth Cary Agassiz, *Louis Agassiz: His Life and Correspondence* (Boston: Houghton Mifflin, 1886), vol. 2, p. 556. That the arrangement of a museum still projects a view of the world is apparent in the vociferousness of the recent controversy in the letters to *Nature* about the reorganization along cladistic lines of displays in the main hall at the Museum of Natural History in London.

8. The sociobiologist Robin Dunbar questions this argument about the transformation of anecdotes into information. He comments: "Isn't it that their function here is to bring alive abstract relationships that have been deduced either from some theoretical consideration or larger body of data? They are not *isolated* examples but *selected* examples."

tive psychologist Frank Beach points out in criticizing such comparisons, a great deal of detail is lost in them, or never found in the first place.[9] What is gained is an analytical category, "graded signals," that can be used to analyze communication in a number of species. The first comparison like this that Wilson presents in the book is apparently intended to be provocative (it is quoted at the beginning of the blurb on the dustjacket): "When the same parameters and quantitative theory are used to analyze both termite colonies and troops of rhesus macaques, we will have a unified science of sociobiology" (p. 4). He carries out this apparently outrageous comparison and then comments, "This comparison may seem facile, but it is out of deliberate oversimplification that the beginnings of a general theory are made" (p. 5). When he later compares the behavior of three species of jays, his methods would be allowed by any zoologist, but when he compares herds of dolphins and ungulates (p. 475), or dinosaurs and elephants as large animals of the plains (p. 446), or castes in ants and vervet monkeys (p. 299), he is making comparisons that some comparative psychologists would consider unreasonable, because the species are so widely separated, and because he is looking more specifically than they would at just a few traits.

Classification is a further step in the stripping away of narrative. If the first thing noticed by the general reader leafing through *Sociobiology* is the pictures, the next thing will be the massive and daunting tables. These tables are not just lists; each supports an argument made in the text, for instance on density-dependent controls (pp. 88–89), chemical communication systems (p. 331), territorial behavior (p. 263), or dominance (p. 292). Each table brings together a number of narratives. For instance, Table 12-1, on "Examples of territorial behavior in which the primary function has been reasonably well established" (pp. 263–64), draws on narratives of the animals encountering other animals, of each animal's life history, of the observer recording these encounters, and of the sociobiologist correlating behavior with functions. The events of these narratives are left out when they are presented in a table, and only return when the information is questioned. When John Krebs criticizes the selection of articles to support Table 12-1, commenting that some of the studies are more reliable than others, he makes the reader reconstruct how each of the functions that were "reasonably well established" were actually established. In the case of another table, when Beach says that infanticide among langurs may be "simply an infrequent, aberrant, and extrane-

9. Beach, "Sociobiology and Interspecific Comparisons," pp. 116–35.

ously induced event" he raises questions about observation and about life histories, calling up all the particulars about who saw what, where, that Table 15-2 (p. 322) cuts out when it lists "infanticide of loser's offspring and insemination by the winning suitor" as one form of sexual selection.[10]

The goal of such classification is not just order; classification is supposed to lead to rules. The tables in the last third of the book attempt to find regular patterns in each order of social animal in the distribution of social behavior with respect to ecology. Sometimes this effort is successful, as in a table based on Jarman's work on ungulates that shows that the social systems "can be transformed with minor distortion into a single axis or sociocline" (p. 479). But for the mammals in general, he finds, "It is difficult, if not impossible, to put this information into one grand evolutionary scheme" (p. 456). For primates he is particularly cautious; he presents two different tables (based on those of Crook and Gartlan and of Eisenberg et al.), arguing that despite what he sees as logical flaws in their arrangements, they have value as analytical tools (see my Epilogue). Ultimately, Wilson wants to make behavioral biology as systematic and quantitative as physics or molecular biology—that is, he wants to remove entirely the narrative elements of particular places, times, and actors. His own ergonomic models of castes (p. 307) are an example of how the natural world can be explained in nonnarrative terms. The various activities of the ants—foraging, fighting, building a nest, laying eggs, caring for the young—become ratios of the weights of castes. Even the individual actors disappear, to be represented by their collective masses. Such models raise one of the key problems for evolutionary narratives, the problem of defining the actor that is the subject of evolution.

The Narratives of Adaptation

If such graphs were the only product of sociobiology, the book might anger some biologists with its methods and criticism, but it would hardly cause a stir outside the discipline. The public controversies concern the larger narratives these tables and graphs serve. Agassiz's synoptic room in his museum would show the glory of God as the creator; Wilson's tables and graphs point toward an equally grand, if rather different narrative, the Darwinian narrative of adaptation. This narrative requires the creation of a subject, and of a species-centered

10. Krebs, "Multiple Review"; Beach, "Sociobiology and Interspecific Comparisons," p. 119.

narrative that holds each of the "Just-So Stories" together, and of a grand evolutionary narrative that structures the whole book, a Great Chain of Behaving. My argument is that *Sociobiology* fills out these very abstract narrative structures with the actors of natural history.

One of the controversial issues in *Sociobiology* is just what the subject of this narrative is. Some critics have accused Wilson of supporting an economy based on competition by emphasizing the individual. But his book is part of a line of thought that seems to eliminate the individual, taking the population as the crucial actor and seeing this population in terms of gene frequencies. In Wilson's definition, "Natural selection is the process whereby certain genes gain representation in the following generations superior to that of other genes located at the same chromosome positions" (p. 3). Wilson's discussions of inclusive fitness, altruism, and group selection have been frequently analyzed elsewhere. What is important to my analysis is the way this construction of a subject both undoes and uses the natural history narrative, with its focus on the individual animal as a character. Wilson quotes Samuel Butler's aphorism, so often cited in biology in the last fifteen years, that "the chicken is only an egg's way of making another egg." The wittiness, the paradox, of this aphorism is in the way it juxtaposes the more familiar narratives in which creatures must be the subjects with the scientific narrative that has genes and populations doing things.

Looking at this view in textual terms, and leaving aside the methodological and philosophical difficulties with such a gene-centered analysis that are emphasized in critiques of sociobiology by geneticists, we can see several possible solutions to the problem of constructing a narrative that apparently lacks a subject. Richard Dawkins worked out a way of telling the story with genes as anthropomorphized characters in *The Selfish Gene.* Another way of telling the story is found in the game-theory work of John Maynard Smith, Geoffrey Parker, and others, who model the organism as if it were a rational strategist. Darwin found a subject by metaphorizing Nature as a careful breeder, in his comparisons of artificial selection with natural selection. Wilson's solution is similar to that of Darwin, but it is perhaps characteristic of the differences between Victorian English culture and American culture today that Wilson compares nature, not to a gentleman farmer, but to an engineer. In discussing animal communication, he says, "If the theory of natural selection is really correct, an evolving species can be metaphorized as a communications engineer who tries to assemble as perfect a transmission device as the materials at hand permit" (p. 240). Later he refers to the "engineering rules" for the evolution of pheromones (p. 370). This language puts Wilson in a long line of writers who treat animals as automata. At the lowest level of organismic response the organism "is like a cheaply-constructed servomechanism" (p. 151). But it must be admitted that he can imagine some very complex automata; he explains the responses of organisms to environmental change in terms of "an immensely complicated multiple tracking system" (p. 145).[11]

This metaphor is worked out for humans in one of the most controversial passages of reductivism, at the end of the book (p. 575).

> The transition from purely phenomenological to fundamental theory in sociology must await a full, neuronal explanation of the human brain. Only when the machinery can be torn down on paper at the level of the cell and put together again will the properties of emotion and ethical judgment become clear. Simulations can then be employed to estimate the full range of behavioral responses and the precision of their homeostatic controls. Stress will be evaluated in terms of neuronal perturbations and their relaxation times. Cognition will be translated into circuitry.

Even Wilson agrees with criticism of this prediction for the future of sociology, but he still argues that the study of human societies will eventually have to go as far as "systems analyses of neuronal populations."[12] When some critics lump Wilson with various behavioral psychologists who actually have quite different views of the causes of behavior, it is perhaps because he shares with them this metaphor of social engineering.

Just-So Stories

Though the metaphor of the engineer will serve to structure opening and closing passages, reminding us of a larger project, it still leaves the problem, in almost every paragraph, of how to tell a story without

11. Robin Dunbar (pers. comm.) emphasizes here that the genes are not themselves actors in this narrative. "The genes are the *currency* of exchange, the individuals are the actors, though it is generally only as *populations* of individuals that Wilson sees them as interesting." He says the difference between Wilson and some other biologists in this emphasis is "due to his being an ant person rather than a mammal person." For an analysis of the "actors" and the "currency" of some accounts of the evolution of society, see Latour and Strum, "Human Social Origins." Ullica Segerstrale comments on the differences between Wilson and some British sociobiologists over such issues as group selection.

12. Wilson, "Multiple Review," p. 717.

a subject. For instance, the section on "The ecology of parental care" begins with an extraordinarily complex and abstract set of narrative chains, of which this is just one (p. 337):

> Expressed in the language of population biology, [the theory] postulates a web of causation leading from a limited set of primary environmental adaptations through alterations in the demographic parameters to the evolution of parental care as a set of enabling devices. The reader can gain the essential idea by studying Figure 16-2. The proposition states that when species adapt to stable, predictable environments, *K* selection tends to prevail over *r* selection, with the following series of demographic consequences that favor the evolution of parental care: the animal will tend to live longer, grow larger, and reproduce at intervals instead of all at once (iteroparity). Further, if the habitat is structured, say, a coral reef as opposed to the open sea, the animal will tend to occupy a home range or territory, or at least return to particular places for feeding and refuge (philopatry). Each of these modifications is best served by the production of a relatively small number of offspring whose survivorship is improved by special attention during their early development.

Note that the "actors" here (that is, the subjects of the action) are *K* selection, demographic consequences, and the production of a small number of offspring. "The animal" mentioned in the middle of the passage is just a counter in the demographics. This bit of narrative is one of the four represented in Figure 16-2 (p. 338), which shows the basic form of the adaptive narrative in four arrows converging from the corners to "Parental Care." Each arrow represents a narrative that begins with an environmental factor, moves through various demographic consequences, and arrives at a change in behavior. The definition that Wilson gives at the beginning of the book describes this same narrative, only in reverse order: "Social evolution is the outcome of the genetic response of populations to ecological pressure within the constraints imposed by phylogenetic criteria" (p. 32).

The empty slot in this abstract narrative of adaptation is filled by bits of natural history. If the reader looks up from the bewilderingly abstract narratives on parental care, like the one I have quoted, he or she sees, on the same page, a very striking picture of a scorpion carrying her young, tiny white miniatures, on her tail. Those who persist past the "language of population biology" to the end of the section come to a charming natural history anecdote about lions teaching their young cubs to hunt (p. 341). These infusions of natural history make the narrative of adaptation come alive. Wilson is careful not to make the animal the conscious subject of evolution—which would make the whole story comic—but his text juxtaposes the two kinds of narrative, and the two scales of the time of evolution and the time of the development of the individual.

The narrative of adaptation has several variants. One can start, not with given environmental factors, as in the parental care stories, but with "phylogenetic inertia," as in the stories of the evolution of social parasitism in ants (p. 364), or of hymenoptera (p. 415), or fly mating (p. 227), or social primates (p. 516). At the level of abstraction of these evolutionary stories, it is sometimes unnecessary or impossible to tell what should come first. A number of the stories allow for multiple paths to the same end, for instance, in the explanations of the relation between environmental unpredictability and species distribution (p. 29), or the evolution of a solitary condition in previously social species (p. 36), or the evolution of sexual dimorphism (p. 334). One would not expect this explanatory flexibility and apparent open-mindedness after reading critics of Wilson's "Just So Stories." But the critics could point out that all the alternative narratives, however different in their implications, follow the same basic story of the adaptiveness of social behavior; in that sense, all the alternatives considered are sociobiological, and explanations from rival disciplinary approaches cannot fit in.

The Great Chain of Behaving

The "Just So Stories" link members of a population syntagmatically from one stage of evolution to the next in almost every section or paragraph of the book. The larger structure of the book, including the order of many of its paragraphs, is given by another narrative of adaptation that links species paradigmatically in a hierarchy. For instance, Wilson ends the important chapter on altruism by asserting that, on the basis of the evidence he has given, "a single strong thread does indeed run from the conduct of termite colonies and turkey brotherhoods to man" (p. 129). It is hard for him, in describing this thread, not to treat the more social species as somehow better, so he uses words like "pinnacles," "haut monde," and "most advanced." But the irony of his grand narrative is that as social evolution is progressing, it is also declining, so that in his terms the most social of all animals are the most primitive (p. 379), the colonial invertebrates. And, not surprisingly, this entomologist finds the ants more social than any mammals. This sort of reasoning parallels that which has traditionally led social philosophers to praise the selflessness of the social insects.

Such a chain of behavior, following the taxonomic chart fom colonial invertebrates up to man, organizes the chapters in the third part of the book. Similarly, many of the sections within chapters in the second part are organized along taxonomic lines, always with the lower orders first and the primates last, for instance in the surveys of play (pp. 165–66), of traditions (pp. 168–72), or of ritualization (pp. 226–28). Of course the chain of behaving is not phylogenetic, and does not imply relations of homology, with humans inheriting a tendency to build cities from the corals. But there is a powerful narrative thrust to these surveys that leads some readers to see Wilson as presenting insect social behavior as the ancestor of human social behavior, instead of as presenting the two kinds of behavior as parallel responses to different environmental challenges.

The controversy about this chain of behaving concerns its end point, with humans. Defenders of Wilson, and sometimes Wilson himself, remind us that man is central in only one chapter of twenty-seven—the last. But links to human behavior are drawn throughout the book, sometimes playfully, often very cautiously, but enough to keep the direction of the narrative clear.[13] A chain that moves up the taxonomic system can be seen structuring one of the most controversial paragraphs of the book, one which asserts that xenophobia can be found in geese, chickens, monkeys, and man.

> The relative calm of a stable dominance hierarchy conceals a potentially violent united front against strangers. The newcomer is a threat to the status of every animal in the group, and he is treated accordingly. Cooperative behavior reaches a peak among the insiders when repelling such an intruder. The sight of an alien bird, for example, energizes a flock of Canada geese, evoking the full panoply of threat displays accompanied by repeated mass approaches and retreats (Klopman, 1968). Chicken farmers are well aware of the

13. The links to man throughout the text include, for example, these passages:
- The defensive array of ungulates is paralleled to Clausewitz's rules of war (p. 45).
- Incest avoidance is linked to the inability of former students to become their teachers' colleagues on equal terms (p. 79).
- Mennonite communites provide an example of the lower limit of herd size (p. 135).
- The waggle dance of bees in compared to Wilson's communication with the reader (p. 177).
- A Harvard commencement is compared to ritualization in birds (p. 224).
- Hormone changes in aggression are illustrated with an example from hockey (p. 253).
- Human occupations are compared to animal roles (p. 313).
- Monkey alloparenting is compared to babysitting (p. 350).

practical implications of xenophobia. A new bird introduced into an organized flock will, unless it is unusually vigorous, suffer attacks for days on end while being forced down to the lowest status. In many cases it will simply expire with little show of resistance. Southwick's experiment (1969), cited in Chapter 11, demonstrated that the appearance of a newcomer is the single most effective means of increasing aggressive behavior in a troop of rhesus monkeys, most of the hostility being directed against the stranger. Human behavior provides some of the best exemplification of the xenophobia principle. Outsiders are almost always a source of tension. If they pose a physical threat, especially to territorial integrity, they loom in our vision as an evil, monolithic force. Efforts are made to reduce them to subhuman status, so that they can be treated without conscience. They are the gooks, the wogs, the krauts, the commies—not like us, another species surely, a force remorselesly dedicated to our destruction, who must be met with equal ruthlessness if we are to survive. Even the gentle bushmen distinguish themselves as the !Kung—the human beings. At this level of "gut feeling," the mental processes of a human being and of a rhesus monkey may well be neurophysiologically homologous.

Parts of this passage are quoted both by Wilson's critics (Alper) and his popularizers (Silcock).[14] The narrative works in two ways. Here Wilson does explicitly say that the rhesus and human behaviors may be homologous, but he is also tracing an analogy (not a homology) from birds to monkeys to man. There is a reference to an experiment, but as with the abstract narrative of adaptation, the story has to be filled in with natural history—the traditional observations of farmers, or the supposed experience of the readers. This paragraph shows the rhetorical shift commented on by so many critics of the last chapter—the slots that in earlier chapters were filled in with observations of animals are filled in chapter 27 with references to common knowledge.

The Territory of the Sociobiologist

The narrative that caused the most antagonistic responses to Wilson's book—the narrative of the future growth of sociobiology itself—seems at first glance not to be related to the narrative of adaptation. This story is told in a way that seems counterproductive. One expects that a

14. Alper, "Ethical and Social Implications," p. 209; Silcock, "How Genetic Is Human Behavior," p. 17.

specialist who wants members of other disciplines to apply the principles of his own discipline will avoid threatening the potential readers he or she seeks to persuade. Niko Tinbergen provides an example of unthreatening rhetoric in a 1968 *Science* article that makes many of the same points Wilson makes in his opening and closing chapters, but apparently does so without arousing antagonism. "As an ethologist," Tinbergen says, "I am going to suggest how my science could assist its sister sciences in their attempts, already well on their way, to make a united, broad fronted, truly biological attack on the problems of behavior."[15] Wilson, on the other hand, uses very aggressive language in dealing with the disciplines nearest sociobiology, whose members he presumably wants to bring over to his vocabulary and methods. A sentence much quoted by his critics says that both ethology and comparative psychology "are destined to be cannibalized by neurophysiology and sensory physiology from one end and sociobiology and behavioral ecology from the other (see Figure 1-2)" (p. 6). Wilson has found the one word most likely to antagonize his readers, and he repeats it at the end of his book when he speaks of neurophysiology *cannibalizing* psychology. Not surprisingly, a number of ethologists and psychologists who reviewed the book sieze on this, the most quoted word in the book, when they criticize what they see as Wilson's misunderstanding of their disciplines.[16]

Whether this aggressive strategy is simply a mistake, or whether it relates to the complex hierarchy of disciplines Wilson lays out in one of his later articles ("Biology and the Social Sciences"), it is consistent with the rest of the book in its emphasis on territory. Several reviewers, sympathetic or unsympathetic, parallel Wilson's view of the sciences to his view of animal competition, referring to the "territoriality" shown in the controversy (not just by Wilson), or to Wilson's assumptions about the "natural selection of academic disciplines."[17] The parallel suggests that the terrain of science is fixed, and resources

15. Tinbergen, "On War and Peace"; Mulkay, Ashmore, and Pinch describe such a project in their unpublished paper on the rhetoric of health economists, "Colonising the Mind" (1986).

16. See, for instance, the comments of the psychologist Frank Beach and the ethologist R. A. Hinde.

17. The phrase is from a review by A. Hunter Dupree. George Barlow ("Multiple Review," p. 701) uses another biological metaphor when he says, "There is an ecology of scientific activity. Valid major ideas, like top-level carnivores, are few. Scientific findings are like primary producers: while numerous and often short-lived, they drive the system." See my paper, "Every Picture Tells a Story," for a comparison of Wilson's diagram of scientific disciplines to a pair of maps of blackbirds' territories.

can be gotten only by defeating a neighboring discipline and usurping its space. A generally favorable reviewer, the ethologist R. A. Hinde, scolds Wilson for his aggressiveness and proposes a less agonistic view of the scientific world: "Whilst Wilson's enthusiasm is infectious, he must not forget that other people are interested in other things."[18] The tone of the reviews suggests that Wilson's disciplinary imperialism had as much to do with the reception of the book as did any of its claims or implications.

Interpretation and Irony in Reviews of Sociobiology

I have argued that what makes *Sociobiology* persuasive is not the facts, not the arguments, but the narrative. I should like to look at some of the many reviews of the book and responses to it as part of a process of interpretation and ironic reinterpretation of this narrative, much as I looked at the published responses to Crews and Fitzgerald's article in chapter 4. There have been a number of different kinds of criticisms, but I shall draw most of my examples from various articles by members of the Sociobiology Study Group (SSG) of Science for the People, which present a case against what they call Wilson's "biological determinism," from Wilson's responses to this group, and from the ensuing exchanges. Other critics—including comparative psychologists, geneticists, anthropologists, and philosophers of science—may not be making the same arguments, but their texts often take similar forms. That is, they do not just disagree with Wilson's arguments, but represent his text so as to weaken his own claims and support theirs. Wilson, in response, uses the same sort of textual devices.

There is a large and still-growing literature on *Sociobiology*, but when one reads the articles one finds many of them are remarkably alike. W. R. Albury comments on this, but is surprised only by the similarity of the various responses to the SSG in defense of *Sociobiology*.

> We have found a high degree of coherence among those responding to the SSG's critique, with regard both to tactics (reversal and reduction) and to strategy (systematic exclusion of politics). It is, of course, possible that this coherence is an artefact of the particular sample of writings studied; but even if this should prove to be so, it

18. Hinde, "Multiple Review," p. 707.

is still significant that such a diverse group of authors should exhibit a unity of this kind.[19]

My contention is that both sides use the same tactics and strategies. Even some Marxist critics practise the "systematic exclusion of politics" from an analysis of the construction of their *own* position. And the processes Albury analyzes as reversal and reduction are, in my terms, part of the reconstruction both of Wilson's and the SSG's texts. I see the process of reconstituting and interpreting Wilson's text in the way reviewers quote him (and he quotes them), the way they place the text in a genealogy (and the way he responds), and in the ways the two sides define the arguments and account for the existence of a controversy.

The focus on texts is itself an indication of a controversy. I have argued in chapter 4 that, although scientific texts are usually treated as transparent, so transparent that they can be summarized sufficiently with a claim and a citation, a controversy makes them opaque, focusing attention on the participants' words and textual strategies. Because, in realist discourse, two incommensurate views of reality cannot both be right, the problem for a realist must be in the formulation or presentation of these views. When the attention is focused on the presentation, it is annoying to those observers who want to find the purely scientific issues in the controversy; Nicholas Wade, writing a news article on the controversy in *Science* complains, "The chief bone of contention . . . thus dissolves into an arid analysis of Wilson's text."[20] I do not find such analysis arid, because I am arguing that the chief bone of contention *is* Wilson's text.

Quoting out of Context

Any review is a reconstruction of the text reviewed. When a review uses quotations, it offers them the way natural history narratives offer facts, as bits of the world that speak for themselves. But the purpose is seldom just to say what those words say; the mere fact of quotation indicates that the writer thinks these words are particularly apt or, more often, particularly and obviously vulnerable.

Wilson charges that he was quoted out of context in some reviews of *Sociobiology*, and his charge is supported by journalists who reported on the controversy for *Science* (Nicholas Wade) and for *New*

19. W. R. Albury, "Politics and Rhetoric in the Sociobiology Debate," *Social Studies of Science* 10 (1980): 532.

20. Wade, "Sociobiology," p. 327.

Scientist (Roger Lewin). But we should remember that all quotation, textually speaking, is out of context. When Wilson quotes a bit of natural history, he omits anything leading up to it and anything after it, perhaps omitting the original author's reasons for making the observation and interpretation of it. Wilson puts it in the context of his own argument, perhaps comparing this behavior, in a way the original author might not accept, to that of another animal in another text. Similarly, when I have quoted Wilson in this chapter, I have not tried to put the quoted passages in the order he put them, and I have often used only a phrase or a sentence from a much longer passage. More important, I have made his words say what I want to say—about the construction of narrative—rather than what he meant to say—about the evolution of social behavior. Other commentators make their own selections for their own reasons. When I read some of the criticisms, especially those of the SSG, I have the sense I am reading a sort of anthology of Wilson's work, for a number of different authors in a number of different articles draw on almost exactly the same quotations, and these quotations all come from just a few sections of the book, especially the first and last chapters and the chapter on altruism. But Wilson himself does such rearranging in his own summary of the book; a quotation he gives when he defends himself is transformed by his highlighting of it, and by his addition of italics to emphasize the point he now wants to make with these words.[21]

Both Wade and Lewin have compared a number of passages quoted in reviews to fuller original versions. I want to look in more detail at one such case to see just what it means when they say Wilson was quoted out of context. After C. H. Waddington's review of *Sociobiology* in the *New York Review of Books*, the *NYRB* published a letter from the SSG, a group consisting mostly of academics from the Boston area, and including some well-known colleagues of Wilson's. The signers disagreed with what they read as a favorable review, and criticized the book for promoting "biological determinism." One full paragraph from the letter reads:

> Another of Wilson's strategies involves a leap of faith from what might be to "what is." For example, as Wilson attempts to shift his arguments smoothly from nonhuman to human behavior, he encounters a factor which differentiates the two: cultural transmission. Of course, Wilson is not unaware of the problem. He presents (p. 550) Dobzhansky's "extreme orthodox view of environmentalism":

21. Wilson, "Multiple Review," p. 698.

> Culture is not inherited through genes; it is acquired by learning from other human beings. . . . In a sense human genes have surrendered their primacy in human evolution to an entirely new non-biological or superorganic agent, culture.
>
> But he ends the paragraph by saying "the very opposite could be true." And suddenly, in the next sentence, the opposite does become true as Wilson calls for "the necessity of anthropological genetics." In other words, we must study the process by which culture is inherited through genes. Thus, it is Wilson's own preference for genetic explanations which is used to persuade the reader to make this jump.[22]

The paragraph uses three bits from Wilson's text (the whole passage from *Sociobiology* is quoted in Appendix 5). The writers are quoting Wilson quoting Dobzhansky, and presumably they quote Wilson's characterization of Dobzhansky's position as the "extreme orthodox view" to show that Wilson is not in his camp. They link his comment that "the opposite could be true" to the quotation by noting that it is at the end of the same paragraph. They present his next sentence as a leap, and show what a leap it is by offering an interpretation of what Wilson means "in other words." The page number they give before their quotation reminds us that these words are in the book for anyone to check; such conventions reinforce the reader's sense that a quotation is a fragment of the original that speaks for itself.

Wilson responds to this charge in another letter to the *New York Review of Books:*

> Allen *et al.* try to make me appear to be the arch hereditarian by quoting my sentence "The very opposite could be true" after a quotation from Dobzhansky stating that "In a sense human genes have surrendered their primacy in human evolution to an entirely new non-biological or superorganic agent, culture." In fact, my sentence came fourteen lines of mostly technical information after the Dobzhansky quotation, and it really followed the sentence "It is not valid to point to the absence of a behavioral trait in one or a few societies as conclusive evidence that the trait is environmentally induced and has no genetic disposition in man." My meaning, which refers to a lesser technical point, was thus grossly distorted by this elision. A reading of the full paragraph will show that I am

22. Allen et al., "Against Sociobiology," p. 262.

far closer to Dobzhansky in my overall view than to the opposite position which seems to be indicated by the mutilated version.[23]

If we look at the paragraph from Wilson quoted in the SSG's letter, we see that his last sentence does comment on a methodological statement, rather than on the statement from Dobzhansky, and that earlier he does in fact imply he is close to Dobzhansky "in his overall view." But he introduces this statement with a subordinate clause, "Although the genes have given away most of their sovereignty," that leads us to expect less than complete agreement with Dobzhansky. And a qualification follows: "they [the genes] maintain a certain amount of influence." He puts the Dobzhansky quotation in his own context, when he makes an apparently parallel statement right after it, so that Dobzhansky's admission that culture is dependent on the human genotype is expanded into the very different statement that genes may influence "the behavioral qualities that underlie variations between cultures." And what Wilson characterizes as "mostly technical information" might be seen as highly controversial support for a position different from the environmentalism of the Dobzhansky quotation, not as data too technical to be considered in the controversy. Also, as the SSG suggests in their ironic quotation of Wilson, his phrase that introduces the quotation, by referring to "the *extreme orthodox* view of environmentalism," implies in the context of scientific rhetoric that such a fixed position should be challenged. If we look at the whole paragraph, we see that while it is indeed supporting some form of environmentalism, in almost every sentence it implicitly questions this position.

If we take just a slightly larger context that the paragraph, looking at the three sentences before it, we find an even more complex interpretation to Wilson's statement that "the very opposite could be true." In these sentences Wilson explicitly states that we should not expect much variation between groups in the human genotype. And he cites, in support of this, a study by Richard Lewontin, his colleague at Harvard, one of his fiercest critics, and one of the authors of the SSG letter. Though Lewontin's work concerned genes specifying blood chemistry, Wilson does not try to argue that the genes he is interested in, those specifying behavior, would be any different. But the apparently gracious last sentence of the paragraph might also be read as a provocation, for it assimilates Lewontin's research to Wilson's argument about social behavior, whereas Lewontin might argue

23. Wilson, "For Sociobiology," p. 266.

that his work has nothing to do with behavior one way or the other, because social behavior is not genetically controlled. His claim, like Dobzhansky's, is supported but is given a quite different interpretation from what he might have intended.[24]

If we look back even further, to the beginning of the paragraph before that quoted by the SSG, we see that Wilson supports his apparent agreement with the environmentalists with some examples from history. Wilson sometimes refers to this passage, in responses to criticism, to show he has always acknowledged that short-term historical change cannot be accounted for genetically. But again the context in which he gives these statements affects our interpretation of them, for he introduces the paragraph by calling this "conventional wisdom." In the rhetoric of scientific texts, such a phrase usually signals an intention to disagree, as the phrase "orthodox view" does in the next paragraph. We would expect Wilson to go on to analyze the deficiencies of the "conventional wisdom."

My purpose here is not to decide if Wilson was or was not unfairly quoted, but to suggest that there is no context large enough to guarantee that a statement will have just one meaning, the intended meaning, that it will speak for itself. In other arguments in the course of the controversy, one side or the other says that the relevant context is to be found in passages in other, less controversial chapters of the book, or in other publications by Wilson, or in earlier publications to which he is responding, or in textbooks encapsulating the assumptions of the discipline, or in comparisons to other contemporary controversies, such as that over research on race and I.Q., or in past controversies such as those over Social Darwinism or immigration, or in the disciplinary history of comparative psychology or ethology. Different kinds of contexts are invoked at different points in the controversy in arguments over the interpretation of Wilson's statements on such issues as territory, sexism, the relations of the social sciences to biology, the relations of modern societies to early human or primate societies, or Wilson's use of anthropomorphic terminology.

For instance, Lawrence Miller argues that the explicit statement by Wilson that sexism is a bad thing should be read in what is presum-

24. Interestingly, nearly every major critic of *Sociobiology* who has done related work, including Lewontin, Levins, Washburn, Beach, and Rosenblatt, is cited prominently and favorably in it. Scientists often tell anecdotes about reviewers responding antagonistically to books that attack or ignore the reviewer's work. But perhaps, in this case, a reviewer is more likely to be antagonized by seeing his or her work appropriated as part of Wilson's argument. Segerstrale mentions this in her account of the relations between Wilson and Lewontin.

ably a larger context, the tendency of the whole book. "We agree with Wilson's *caveat:* sexism is not justifiable. But the thrust of Wilson's point, here and elsewhere, is that sexism is inevitable, even if undesirable, because it is genetically determined." Here the relevant context, the *thrust* of Wilson's text, is something separate from any quotable statement in it. Joseph Alper transforms Wilson's many statements on population "groups" into statements on "race" by noting that, although Wilson is careful not to use the word "race" and although "group" does not usually mean "race" in the discourse of evolutionary biology, another author, R. A. Goldsby, *has* given a definition of "race" such that one can say "race, as defined by Goldsby, is one type of Wilsonian group."[25] In this rather complicated process of translation, the context for determining the meaning of the author's statements is not in his text, or even in the disciplinary discourse of which it is a part, but is in another text that gives the real meaning of his terms, such that he can be seen to say something even if he deliberately does not say it.

Wilson's characteristic style may make it easier to pick out damning quotations than it is with some scientists; he tends to intersperse long stretches of cautious suggestion and qualification with a few brash overstatements. But we can see such selective quotations in many scientific controversies, and indeed we can see it on both sides of this controversy. Members of the Sociobiology Study Group make the same sort of claims that they were misinterpreted that Wilson makes about their reading of his text. Responding to an analysis of the controversy by Arthur Caplan, Lawrence Miller says, "the article represents a systematic misunderstanding of our critique of sociobiology." Responding to the *Science* article, Joseph Alper and his colleagues say "Wade distorts and, in effect, trivializes the whole matter." Steven Rose says, after the article in *New Scientist,* "Roger Lewin's account of the sociobiology controversy . . . suffers, it seems to me, from just those vices he accuses E. O. Wilson's critics of adopting—selective quotation to distort the nature of the charges being made."[26] My point is that Lewin does use selective quotation and paraphrase to put statements in a context other than that intended, but so does Rose in response to him, and so does another letter in response to Rose's

25. Miller, "Philosophy, Dichotomies, and Sociobiology," p. 322; Alper, "Ethical and Social Implications," p. 208.

26. Miller, "Philosophy, Dichotomies, and Sociobiology," p. 319; Alper et al., "The Implications of Sociobiology" (see also Alper, "Ethical and Social Implications," p. 206); Rose, "Sociobiology," p. 433.

letter (Hammerton), and so does Rose again in the response to Hammerton's letter ("Sociopolitics"). And so do I, in presenting this sequence. All participants in the controversy assume that there is a self-evident meaning to their words, and that the reader who wants to find what Wilson calls "my true meaning" or what Rose calls "what Wilson actually does say" need only go to the trouble of looking up the original text. I am arguing that the various contexts introduced by the participants themselves make it impossible to settle on one meaning on the basis of the text alone. A discourse community can settle on one interpretation of a text, and I think both sociobiologists on the one hand, and critics of sociobiology on the other have reached consensus about how to read *Sociobiology.* But such agreement comes at the end of a controversy, through the controversy, not at the beginning. The assertion that texts have perfectly straightforward, context-free meanings that can be found by application of rational rules is just another rhetorical tool, and it is a tool equally available to both sides.

In all these cases, the charge of misinterpretation is used to rebut criticism. References to misinterpretation can also serve as a polite cover for disagreement, one that does not directly challenge the competence of the scientist criticized. For instance, reviewers who essentially agree with Wilson focus on issues of interpretation when they want to point out some issue on which they differ from him. The explanation of the difference cannot be in nature, because nature can have only one correct interpretation, and it cannot be due to an error on Wilson's part, or one's own part, so it must be due to misreading. This is especially true in the generally favorable comments by other behavioral biologists in a collective review in *Animal Behaviour.* G. P. Baerends, after praising *Sociobiology,* accounts for his differences with Wilson as errors in Wilson's reading: "I find Wilson's discussion of Tinbergen's conflict hypothesis . . . based on insufficient and biased use of the available evidence." In his sympathetic review John Krebs comments, "I feel that in some places Wilson is not discriminating enough in his literature reviews. In summary tables such as 12-1 . . . he makes no distinction between short-term observational studies published in popular magazines and painstaking experimental work done over several years."

Jerry Hirsch's attack in this same collective review, though remarkably vicious and personal, uses the same device used by those who generally agree with Wilson, to end his comment on a note of apparent concern and respect for Wilson's abilities: "This has been done in the hopes of encouraging Wilson to read more carefully, to prune his bibliography of trendy sources (*Atlantic Monthly, Time, Scientific Ameri-*

can, etc.; and also unreliable claims in the primary journals)."[27] It is clear from Hirsch's comments that he and Wilson differ on a number of fundamental methodological and social issues, but for the purposes of a disinterested conclusion, these can all be reduced to the advice to "read more carefully." For these purposes, Hirsch takes the reading of the primary journals, and the weeding out of "unreliable claims," to be unproblematic. But in their exchange in *Animal Behaviour,* Wilson and Hirsch disagree about just what is claimed in an article by Dobzhansky. Hirsch reprints apparently inconclusive figures to show that the article can offer no support for Wilson's claim of experimentally demonstrated rapid speciation, whereas Wilson says, "the unevenness in the progress of selection stressed by Hirsch was explained by Dobzhansky and Pavlovsky as largely an artifact having no bearing on the main result." That they can differ so completely shows again how the same quoted words, and even the same figures, can have radically different meanings when placed in different contexts.

Tracing the Genealogy

One's interpretation of the book depends not only on the context in which one reads passages word by word, but also on the context in which one puts the book as a whole, and the selection of this context is another interpretive decision in which one can ironically invert the apparent narrative. We saw in chapter 4 that participants in that controversy within the core set gave sharply contrasting introductory reviews of the literature. Almost every reviewer of *Sociobiology* gives it a genealogy, a set of texts against which it is to be read. Two sorts of genealogies figure frequently in the controversy, and both genealogies are open to reinterpretation by critics or defenders of the book.[28]

J. R. Krebs's brief review opens with a passage that can be taken as typical of those in reviews by other ethologists. He locates the book in disciplinary terms (Krebs is an ethologist), as part of a personal tradition of major figures, in terms of national traditions (Krebs is British), and in contrast to earlier popularized treatments of sociobiological concepts.

> A biochemist recently asked me to define sociobiology. The only simple answer to the question was "The branch of biology covered

27. Baerends, "Multiple Review," p. 700; Krebs, "Multiple Review," p. 709; Hirsch, "Multiple Review," p. 709.

28. I discuss cartoons and visual suggestions of a genealogy in "Every Picture Tells a Story."

> by E. O. Wilson's book." There is no other definitive work on this eclectic field which combines behavioral ecology, population biology, and evolutionary theory. Although many of the earlier key developments in sociobiology were due to European workers such as Crook, W. D. Hamilton, Maynard Smith, and the followers of Lack and Tinbergen, the mainstream of European ethology sadly failed to follow up the lead, so that most of the more exciting recent developments have come from North America. R. A. Hinde's seminal work, *Animal Behaviour,* makes no mention of Hamilton and Maynard Smith, and only brief reference to the work of Lack and Crook. It goes without saying that Wilson's book is outstanding, both as an encyclopedic review of the literature and as a lucid, critical synthesis of theoretical concepts.

Krebs credits Wilson with defining a field, created by combining three separate fields in the "synthesis" referred to in Wilson's title, but it is a field that already has some distinguished members who worked under other disciplinary titles. He puts Wilson in a tradition with several British researchers, some of whose names appear at the start of most favorable reviews, and compares the book favorably with that of the best known British ethologist. When he says the quality of the book "goes without saying," he acknowledges, as do most of the favorable reviews, Wilson's considerable reputation from his earlier books. The last sentence of Krebs's review also helps place Wilson, by contrasting his work with that of the authors of earlier sociobiological bestsellers: "Those who accepted uncritically the views of the evolution of social behavior popularized by Lorenz, Ardrey, and Tiger should study E. O. Wilson to learn the proper version of the story."[29]

In the opening of the NYRB letter criticizing the book, the SSG give it a different genealogy, placing it in terms of past figures who held views that are said to be analogous, and in terms of its possible future social effects.

> Beginning with Darwin's theories of natural selection 125 years ago, new biological and genetic information has played a significant role in the development of social and political policy. From Herbert Spencer, who coined the phrase "survival of the fittest," to Konrad Lorenz, Robert Ardrey, and now E. O. Wilson, we have seen proclaimed the primacy of natural selection in determining the most important characteristics of human behavior. These theo-

29. Krebs, "Multiple Review," pp. 709, 710.

> ries have resulted in a deterministic view of human societies and human action. Another form of this "biological determinism" appears in the claim that genetic theory and data can explain the origin of certain social problems, e.g., the suggestion by eugenicists such as Davenport in the early twentieth century that a host of examples of "deviant" behavior—criminality, alcoholism, etc.—are genetically based; or the more recent claims for a genetic basis of racial differences in intelligence by Arthur Jensen, William Shockley, and others.[30]

Here Wilson figures, not as the first synthesizer of several scientific fields, but as the latest in a line of scientists who applied biology to social policy. The theories of these scientists have apparently led people to have a "deterministic view," and their theories are related to, but are apparently not the same thing as, those of another and more virulent line that leads to the two best known academic proponents of racial differences in intelligence. After a paragraph accounting for the persistence of these wrong ideas, the writers show how dangerous they are by noting the use of Social Darwinism by J. D. Rockerfeller, Sr., to justify his practices, and by tracing American racism and German Fascism to eugenics: "These theories provided an important basis for the enactment of sterilization laws and restrictive immigration laws by the United States between 1910 and 1930 and also for the eugenics policies which led to the establishment of gas chambers in Nazi Germany."

Although the political genealogy does not actually contradict the scientific genealogy, no reviewer places Wilson in both contexts. He must be either the product of scientific progress or the product of ideological reproduction. But it is possible to alter either the scientific or the political genealogy so that its significance is reversed. For instance, the familiar geographical terms of Krebs's scientific genealogy—the Europeans failing to follow up their lead and the Americans taking over—can be rearranged in terms less favorable to Wilson. The American anthropologist S. L. Washburn includes Wilson in the tradition of "European thinking," condemning it for "eugenics, racist theories" and other errors. But the the *NYRB* review by the British evolutionary biologist C. H. Waddington (the review that ostensibly inspired the SSG letter) criticizes Wilson by associating him with "certain algebraic theories about population growth recently developed by American authors" which "biologists on the other side of the Atlantic feel . . . are

30. Allen et al., "Against Sociobiology," pp. 259–60.

rather schematic, and never fully apply to the complicated situations that arise in the actual ecological situations of nature."[31] Since Wilson is an American who reads British and continental scientists, he can be placed intellectually on either side of the Atlantic, and his critics, British or American, prefer to put him on the other side.

The revision of the genealogy can also rearrange the line of scientific figures from which Wilson derives. So, for instance, a comparative psychologist like Ethel Tobach removes Wilson from the central scientific position Krebs gives him, relating him only to ethology and contrasting this tradition with the comparative psychology tradition of C. L. Morgan, T. C. Schneirla, and D. E. Lehrman.[32] Also, she introduces other figures who said what Wilson said before Wilson said it, so that Wilson's book is not new—the genealogy then becomes a priority account. Almost every figure in Krebs's genealogy is rescued from this association with Wilson by one or another reviewer who uses this model figure—whether it is Tinbergen, or Crook, or Maynard Smith, or Hamilton—to show how better scientists avoided Wilson's errors. Stephen Jay Gould elegantly opens an essay on sociobiology by praising Linnaeus. Although most critics associate Wilson with Konrad Lorenz to discredit Wilson (political critics often cite a racist essay Lorenz wrote in 1940), Mary Midgely uses Lorenz's *Behind the Mirror* as "the proper guidebook" that shows in contrast the weakness of Wilson's philosophical position ("Rival Fatalisms"). A key figure in these lists of names is J. B. S. Haldane; British sociobiologists claim him as a founder of mathematical evolutionary genetics, whereas Marxist scientists look back to him as a figure who could combine scientific work with political action.[33]

Wilson, of course, tries to disassociate himself from a political genealogy that leads back to J. D. Rockefeller and Hitler. He argues that sociobiology, by showing altruism to be adaptive, actually refutes

31. Washburn, "Animal Behavior and Social Anthropology," p. 63; Waddington, "Mindless Societies," p. 256.

32. Tobach, "Multiple Review." David Crews points out that Tobach and Wilson reenact the old debate between their mentors Schneirla and Wheeler.

33. See Gary Werskey, *The Visible College* (London: Allen Lane, 1978), on Haldane's politics, and Ronald Clark for a biography, *J.B.S.: The Life and Work of J.B.S. Haldane* (London: Hodder and Stoughton, 1968); it unfortunately skimps on scientific detail. Haldane is another remarkable popularizer whose style combines cautious science with brash pronouncements. One can imagine a selection of statements from his writings that could make him seem politically reactionary; in fact, one can imagine a selection that could make him seem to say practically anything. In the "Preface" to his collection of popular essays, *The Inequality of Man* (Harmondsworth: Penguin, 1937), he invents an amusing example of how he could be misinterpreted in the press.

Social Darwinism, with its emphasis on individual fitness. He specifically attacks the views of Shockley and of earlier eugenicists. He argues that the method of "strong inference" separates him from popularizers who use the "advocacy method." He seldom cites Lorenz except to attack his assumptions and models as primitive or simplistic, and he singles out Lorenz's popular and controversial book *On Aggression* for attack. Wilson rejects any genealogy that puts him in a line of social engineers; he often argues that to treat his descriptions of social behavior as prescriptions for social policy is to commit the "naturalistic fallacy," confusing what is with what ought to be.

But Wilson's most effective response to the political genealogy is to turn the accusations against him back against his accusers, to use their words ironically as they use his words. This tactic is analyzed well by Albury, but Albury does not note that the SSG uses the same sorts of ironic reversal. When Wilson quotes the reference in the *NYRB* letter to Nazi gas chambers, it is not, presumably, to give the charge wider circulation, but to show that his critics have gone over the top and proved themselves to be "political" rather than scientific. The review by Jerry Hirsch, to which I already referred, repeats the affiliation of Wilson with Shockley in this form: "He [Wilson] was once bothered enough about heritability . . . to send me his manuscript in advance and then, like William B. Shockley, to telephone long distance in an unsuccessful attempt to recruit my support." Wilson responds by using the tenuousness of this particular attempt at affiliation to undermine Hirsch's whole review: "To connect my name with that of William Shockley, a notorious professed racist, on the sole ground that we both talked about heritability on the telephone to Hirsch, is in my opinion a tactic that should be beneath the reviewer of a scholarly work. I can only interpret it as further evidence that in this particular review political criteria were used to judge science."[34]

This ironic turn is a potent response. Later criticisms respond to Wilson's attempts to separate himself from any political genealogy. Steven Rose rebuts the charge that Wilson is being persecuted like earlier scientific heroes: "Far from being much abused 'new Galileos,' as their advocates have claimed for them, the sociobiologists are mere Ptolemaic medieval schoolmen." Rose's response to Wilson's response is again ironic, echoing the text of his opponent—in this case, Wilson's supporters' protests of persecution—to give it a meaning exactly the opposite of its apparent meaning. The critics respond to Wilson's declarations of political liberalism, his denials of any attempt

34. Hirsch, "Multiple Review," p. 708; Wilson, "Multiple Review," p. 718.

to perpetuate the oppression of women or blacks, by insisting that they are tracing a line of ideas and effects, so that Wilson's own political statements are irrelevant to the effect of his ideas. The Sociobiology Study Group says, in one of its later articles, that "such determinism provides a direct justification for the status quo as 'natural,' although some determinists dissociate themselves from some of the consequences of their arguments." Still, the critics do search Wilson's writings for some explicit and incriminating political statement. One such statement is found by Rose, Lewontin, and Kamin in an interview in *Le Monde* in which Wilson, "identified himself with American neoconservative libertarianism."[35] That they would find this statement in such an out-of-the-way source, rather than in any of Wilson's many, many writings or his interviews in English, suggests both that Wilson avoids making any explicit political statement and that his critics, despite their larger critique of ideology that makes such self-declarations irrelevant, need him to make such statements. The critics have an easier time finding such statements in the less cautious prose of Richard Dawkins, and so they often link his book *The Selfish Gene* to *Sociobiology,* though the two texts are quite different, the two approaches have some important theoretical differences, and the authors do not refer to each other.

These genealogies change in the course of the controversy; Albury points out that the political genealogy was often softened for rhetorical reasons in the later versions. The supporters who offered the professional genealogy revise it somewhat in retrospect, and Wilson's later texts reinterpret his stance in *Sociobiology.* Those who criticize *Sociobiology* often attack it by using quotations from Wilson's later books, so that *Sociobiology* is reread in terms of *On Human Nature.* Rose, Lewontin, and Kamin say, "The development of the literature of sociobiology since 1975 . . . including Wilson's own *On Human Nature,* leave little doubt that the problem of human nature is at the center of sociobiological concerns." This assumes that the "literature of sociobiology" is whatever Wilson writes; if one looked for this literature in the writings of Maynard Smith, Parker, and other British sociobiologists, one might well doubt that "human nature" was the central problem. The critics of *Sociobiology* might want to criticize this work too; my point is that they do not need to deal with the huge literature of the field in the last ten years as long as they can define "the literature" as being synonymous with Wilson's later writings and

35. Rose, "It's Only Human Nature"; SSG, "Sociobiology," p. 280; Rose et al., *Not in Our Genes,* p. 264.

with the more extravagant sociological and anthropological claims based on them.[36]

Retrospectives by some of these British researchers, written after the controversy died down somewhat, have made a different sort of revision of Wilson's genealogy. Geoff Parker, Maynard Smith, and J. R. Krebs have stressed more the importance of the work leading up to Wilson's book, and have in one way or another distanced themselves from his methods of argument and his conclusions about human nature. For instance, Krebs, whose genealogy I have quoted, credited Wilson in his 1975 review with defining the field of sociobiology. But in his 1985 retrospective he needs to stress that the book merely coincided with the publication of major work by other people. "E. O. Wilson's book *Sociobiology*, published ten years ago, was by no means the start of sociobiology and behavioural ecology. But its publication coincided with, and perhaps helped to sustain, an almost explosive growth of interest in the subject." Whereas earlier Krebs had praised Wilson's work as "outstanding, both as an encyclopedic review of the literature and as a lucid, critical synthesis of theoretical concepts," his later review says, "It is an eclectic compendium of facts with a limited amount of theorizing." The retrospective is still favorable, as are all the recent retrospectives by evolutionary biologists that I have seen. But one can sense in it that the controversy over the book was not entirely appreciated by others who had long been working in fields that have come to be defined as sociobiology. For instance, in 1982, Patrick Bateson explicitly disassociated researchers in the King's College Sociobiology Group from the more controversial aspects of sociobiology: "Nobody who knew their work could accuse them of doing bad science. Furthermore, they would tolerate neither sloppy argument nor extravagant generalisations from studies of animals to humans."[37] To return to the metaphor I have used, in which texts are collected like items in a museum, the response of other sociobiologists to Wilson can be likened to the response of the curator of a natural history museum who arrives at work to find a living wooly mammoth in the great hall; it brings in the crowds, it's worth studying, and it certainly deserves respect, but every time it moves it messes up all the other cases.

36. Rose et al., *Not in Our Genes,* p. 243; see Gould, "Biological Potential," for a criticism of an article containing the most absurd of such claims.

37. Krebs, "Sociobiology Ten Years On," p. 40; Bateson, "Preface," p. x.

The Truth Lies Somewhere in Between

In quoting and paraphrasing, writers on both sides of the controversy create a version of the texts they read and criticize. In giving genealogies, writers create a version of the text's past and future affiliations. Writers also give a version of the controversy, a narrative that explains how there could be disagreements about matters of fact. Two devices particularly interest me in these versions: the creation of opposition, and the asymmetrical accounting for controversy. That is, various participants define the issues of the controversy in terms of two poles, and then place themselves between the two poles while their opponents are at one or the other of the extremes. And various participants explain why there is a controversy by giving a sociological explanation for their opponents' errors, while explaining their own position in scientific terms.

The creation of opposition has been a frequently used device, not only in the sociobiology controversy, but in earlier controversies on similar issues. In the kind of heated debate that Nature vs. Nurture always involves, any participant can step in as the voice of reason. For instance, Niko Tinbergen builds his 1968 *Science* essay around the apparent opposition between ethology and comparative psychology, Lorenz and Schneirla, which he can then reconcile with his own more moderate reading of both positions. Wilson devotes part of an introductory chapter of *Sociobiology* to listing the various "Dualities of Evolutionary Biology," many of which crumble into "semantic ambiguity" upon his closer examination. He uses the device of oppositions in the passage I have discussed when he defines his apparently moderate position against the "extreme environmentalism" described in the quotation from Dobzhansky.

The Sociobiology Study Group are equally careful not to be identified with an extreme position.

> We are not denying that there are genetic components to human behavior. But we suspect that human biological universals are to be discovered more in the generalities of eating, excreting, and sleeping than in such specific and highly variable habits as warfare, sexual exploitation of women and use of money as a medium of exchange.

Wilson responds to the SSG's criticism of his "biological determinism" by proposing two poles and placing himself between them, so that the SSG position, and not his own, is seen as the extreme:

> In their earlier *New York Review* statement (Allen et al. 1975) the group . . . maintained that although eating, excreting, and sleeping may be genetically determined, social behavior is entirely learned; this belief has been developed further in the *BioScience* article. In contrast, and regardless of all they have said, I am ideologically indifferent to the degree of determinism in human behavior. If human beings proved infinitely malleable, as they hope, then one could justify any social or economic arrangement according to his personal value system. If on the other hand, human beings proved completely fixed, then the status quo could be justified as unavoidable.
>
> Few reasonable persons take the first extreme position and none the second. On the basis of objective evidence the truth appears to lie somewhere in between, closer to the environmentalist than to the genetic pole. That was my wholly empirical conclusion in *Sociobiology: The New Synthesis* and continues to be in my later writings.

Wilson accuses the SSG of misrepresenting his views "in order to have a conspicuous straw man."[38] But the strategy on both sides actually requires two opposing straw men. Mary Midgely exploits this symmetry in her philosophical analysis, "Rival Fatalisms: The Hollowness of the Sociobiology Debate." What she misses, I think, when she ends by taking position between the two extremes (a moderate position she identifies with Lorenz and Irenäus Eibl-Eibesfeldt) is that all the other participants can make the same rhetorical move. Compare the passage I've just quoted, in which Wilson rejects both extreme positions and places himself closer to environmentalism, to this passage from Steven Rose's critique:

> The real failure of the sociobiologists lies in their seeming inability to avoid the either/or trap. Behavior must be either socially or biologically determined, or must represent the arithmetic sum of a biological (genetic) and an environmental component. On the contrary, a proper understanding of the interaction of the biological and the social in the production of humans and their society will only be possible following the simple recognition that both genes and environment are perfectly necessary to the expression of any behavior.[39]

38. Allen et al., "Against Sociobiology," p. 264; Wilson, "Academic Vigilantism," pp. 292, 293.

39. Rose, "It's Only Human Nature," p. 169.

It is hard to imagine Wilson, or any other biologist, disagreeing with Rose's position as it is stated in the last sentence. Indeed, on the evidence of all these passages, a literal-minded reader might think that all the participants in the debate, without their realizing it, are in agreement. Of course they are not; they are agreed only in using the same device, defining opposing positions by presenting them as extreme poles of nature or nurture.

Call in the Sociologists

Part of the controversy is in accounting for why there is a controversy in the first place. The controversy is presented by all participants as being in some way unpleasant and unscientific. "Beneath the smoke," says Wade "is a scientific issue." Participants on both sides see social forces at work in the scientific controversy, but both invoke them only to explain false positions—the sociologists are called in to talk to the other guy. Richard Burian, a critic of sociobiology, says, "It is of no little sociological interest that sociobiology has met with considerable success in the academic world; publishers, universities, professional associations, and many working scientists believe that it has already established itself as a legitimate scientific discipline." Albury sees the presence of "a sociopolitical element" as a problem for "the defenders of sociobiology and their philosophical allies," but does not seem to think it a problem for the scientists criticizing sociobiology. On the other hand, Gerald Holton quotes the comments of Alexander Morin of the NSF, accounting for the opposition to sociobiology in theological terms: "Why does it arouse such passionate opposition, even among people who, in other fields of enquiry, are (or appear to be) dispassionate in their scientific consideration of science? Because what we are seeing, I think, is not a scientific response to evidence but a doctrinal response to heresy."[40] Holton's analysis, too, takes the scientific response to evidence for granted, as undiscussable, while calling for the social analysis of the unscientific. In Burian's case, the anomaly is the social success of a pseudoscience; In Morin's case, the analysis is to be made in the familiar terms of religion versus science. These two kinds of explanation of controversy, one based on social factors and the other on psychological, are similar in being applied asymmetrically. Good science apparently needs no explanation.

A number of accounts explain the controversy by saying that the other side has an unscientific emotional interest in the debate. The

40. Wade, "Sociobiology," p. 325; Burian, "Methodological Critique," p. 392; Albury, "Politics and Rhetoric," p. 529; Holton, "The New Synthesis," p. 82n.

SSG's letter in the *NYRB* says, "What Wilson's book illustrates to us is the enormous difficulty in separating out not only the effects of the environment (e.g., cultural transmission) but also the personal and class prejudices of the researcher." In his response, Wilson comments on the "remorseless zeal" of the SSG. Arthur Caplan refers to the SSG's letter as "impassioned" and to Wilson's response as "equally impassioned." The geneticist James King says, of controversies about field studies of animals, that "The intense emotional involvement of sociobiologists makes it hard to achieve scientific give and take with them." But a sociobiologist might detect signs of King's own emotional involvement, when, in the conclusion to his article, he says that in sociobiology "a jerry-built doctrine has been compounded of old hat genetics that current research has already rendered obsolete, of sophomorically cynical interpretations of social relations, and of a doctrinaire rejection of the contribution of ontogeny to the behavioral phenotype." This is not an invitation to scientific give and take.[41]

Burian's reference to the academic success of sociobiology is an example of another sort of attribution of unscientific, socially contingent influences; both sides argue that the other side is accepted merely because it is fashionable. Washburn regrets that it is "fashionable to minimize the nature–nurture argument" as Wilson does in his introduction. Waddington calls altruism "a fashionable topic for a rather foolish controversy." Steven Rose says, of attempts to account for social change in sociobiological terms, "at best the exercise becomes a piece of fashionable Harvard or Oxford intellectual games-playing; at worst a way of ideologically justifying the status quo."[42]

The suspicion of the fashionable remains even when the writer is not entirely hostile to the fashionable idea. In one of the first reviews, Donald Stone Sade expresses his skepticism about the manner of proponents of inclusive fitness and kin selection. "In the conference chambers of scientific meetings I have seen these ideas, like the sweet smoke of a forbidden weed, create a sense of euphoria among their advocates, who seem on the verge of some hidden truth, obscure until the inhalation of these heady notions. Wilson, by contrast, appears to intend his book to be a challenge." Sade and other supporters of Wilson use this device to distinguish him from less scholarly writers. Mary Midgely considers Wilson's followers to be more dangerous

41. Allen et al., "Against Sociobiology," p. 264; Wilson, "Academic Vigilantism," p. 298; Caplan, "Ethics," p. 309; King, "Genetics of Sociobiology," pp. 101, 104.

42. Washburn, "Animal Behavior," p. 57; Waddington, "Mindless Societies," p. 254; Rose, "It's Only Human Nature," p. 167.

than Wilson himself, because their convictions, unlike his, are based on the blind following of fashion. "Like any flag-waving movement, as it gathers strength it is bound to attract a mass of supporters who will catch their leaders' confidence without his scruples and without understanding his limitations . . . The academic world is full of people who ask nothing better than to settle into such an army. Wilson is a prophet, and he isn't going to lack acolytes." Of course, the same phenomenon of a new figure rapidly gathering support can be described as "consensus" or as "broad agreement" or as "a new paradigm" if it supports one's own position.[43] (And as we saw in the biologists' comments in ten-year retrospectives, there seems to be, if anything, a tendency for those whose ideas are closest to those of the controversial figure to try to distance themselves from him, not to appear as his acolytes.)

One version of this explanation in terms of fashion associates the view one opposes with popularity outside the scientific community. Wilson himself uses this popularity against the earlier popularizing sociobiologists, and his supporters contrast his difficult and scholarly work with the bestsellers. On the other hand, almost any essay critical of Wilson, by the Sociobiology Study Group, Gould, Rose, Montague, or others, starts with the popular success of *Sociobiology* and the appearance of its argument in such nonscientific magazines as *House and Garden* and *People.* We have seen that Jerry Hirsch criticizes Wilson for including citations to such journals as *Atlantic* and *Scientific American.* In each case, public interest and acceptance is itself evidence of the unscientific nature of the argument. It may seem odd to find this tactic being used by authors like Gould, Rose, and Wilson himself, who are highly successful popularizers as well as prominent researchers. But as we saw in chapter 5, there is a curious ambivalence in analyses of the process of popularization; it can be seen either as the vulgarization of pure science to pander to the tastes of the ignorant, or as the stripping away of the obfuscation of the specialists by talented writers who can make essential ideas accessible. So here the attempts to reach a wide audience with one's claims can be seen either as a laudable awareness of the social implications of scientific ideas, or as a dangerous attempt to enlist the authority of science for one's personal political beliefs.

43. Sade, "Evolution of Sociality," p. 244; Midgley, "Rival Fatalisms," p. 26. Nigel Gilbert and Michael Mulkay analyze one case of the representation of consensus in *Opening Pandora's Box,* (pp. 112–40).

Methodological Parodies

It might seem that the controversy would come down to questions of methodology and philosophies of science. And since most participants agree in using the vocabulary of Karl Popper (those who make their case based on Kuhn or Paul Feyerabend are looked upon with suspicion by both sides) it might seem that there were grounds for distinguishing scientific from nonscientific practice. But as in other controversies, there is disagreement, especially between practitioners of various disciplines, about what counts as a falsifiable hypothesis and what counts as a test. The language of Popper becomes a general rhetorical resource that has little to do with the structures of the texts.[44]

The most common form of methodological argument in the controversy is to present a parody of the methods given by the opposing side. For instance, both the Sociobiology Study Group and Rose, Lewontin, and Kamin structure their attack around a story of how the sociobiologist proceeds. They present sociobiological method as a "Just-So Story" in which present-day features are reified and then projected into the past to give a pseudohistorical explanation.

> Sociobiology, as a theory of human society, is built of three parts. First, there is a description of the phenomenon it is meant to explain, that is, a statement of human nature. This description consists of an extensive list of characteristics that are thought to be universals in human societies, including such diverse phenomena as athletics, dancing, cooking, religion, territoriality, entepreneurship, xenophobia, warfare, and the female orgasm.
>
> Second, having described human nature, sociobiologists claim that the universal characteristics are encoded in the human genotype. . . .
>
> The third step in the sociobiological argument is the attempt to establish that the genetically-based human universals have been established by natural selection during the course of human biological evolution. . . .

44. See Jonathan Potter, "Testability, Flexibility; Kuhnian Values in Scientists' Discourse Concerning Theory Choice," *Philosophy of the Social Sciences* 14 (1984): 303–30, for a study of the use of this rhetoric by psychologists.

> In what follows, we look more closely at these three elements of sociobiology: the description of human nature, the claim of its innateness, and the argument for its adaptive origin.[45]

This version simply reverses the adaptive narrative that Wilson uses in his book, so that now the causes run from the trait, through demography, to the environment, and the explanations move from humans to animals rather than from animals to humans. The force of such a parody comes, not from comparison to good science, to the authors' own scientific work, or to some philosophical model, but from the ironic reversal of Wilson's own methodological model.

In *Sociobiology,* Wilson's methodological criticisms, and his methodological parodies, are reserved for researchers in the disciplines closest to his own; he uses them to suggest the strength of this own method by contrast. For instance, he criticizes the best-selling books on sociobiology that preceded his for using what he calls "the advocacy method" (p. 28).

> In sociobiology, it is still considered respectable to use what might be called the advocacy method of developing science. Author X proposes a hypothesis to account for a certain phenomenon, selecting and arranging his evidence in the most persuasive manner possible. Author Y then rebuts X in part or in whole, raising a second hypothesis and arguing his case with equal conviction. Verbal skill now becomes a significant factor. Perhaps at this stage author Z appears as an *amicus curiae,* siding with one or the other or concluding that both have pieces of the truth that can be put together to form a third hypothesis—and so forth seriatim through many journals and over years of time.

This often-quoted passage comes early in the book. Wilson goes on to present the correct method for pursuing sociobiology, through "strong inference." This method, though attractive, is both unconvincing and unwieldy, and Wilson refers to it again only in token passages through the book. It is not his use of strong inference, but his irony in

45. The quotation is from Rose et al., *Not in Our Genes,* p. 243; SSG, "Sociobiology," p. 282.

describing the advocacy method that makes us see him as doing something different.[46]

Wilson also gave methodological accounts of the various disciplines that are to go into the sociobiological synthesis. These provoked angry responses from anthropologists, geneticists, and comparative psychologists, despite, or perhaps because of, the apparent respect Wilson has for their findings. The following description of the work of some comparative psychologists can serve as an example of Wilson's double-edged prose which, without exactly criticizing, produces a description that I imagine would not be accepted by comparative psychologists (p. 349).

> Although vertebrate studies are marked by eclecticism, as Lehrman and Rosenblatt said, much of the work remains motivated by a very few strong, albeit implicit themes. One is environmentalism. The background of a majority of the researchers is in anthropology or experimental psychology, in which there exists a bias to assign as much of the measured intraspecific variance of behavioral traits as possible to environmental influences. There is nothing wrong with this attitude; it can be quite heuristic as long as it is kept explicit. The bias results in a determined probe to catalogue and weigh all possible environmental factors, both those manifest in naturalistic studies of free-ranging populations and others that become apparent only when their effects are magnified through experimental manipulation. . . . The developmental psychologists cannot be too

46. The irony of Wilson's paraphrase and commentary also serves to distance him from these methods of John C. Lilly's popular books on dolphins, which he fears have misled the public and other scientists about the methods of sociobiology. I shall quote enough to give a sense of the tone (p. 474):

> Although Lilly never states flatly that the dolphins and other dolphinids are the alien intelligences he seeks, he constantly implies it. . . .
> Anecdotes are used to launch sweeping speculations. . . .
> This fantasy is then turned into a premise for even stronger discussion and speculation. . . .
> This example fairly represents the overall quality of Lilly's documentation and logic. Objective studies of behavior under natural conditions are missing, while "experiments" purporting to demonstrate higher intelligence consist mostly of anecdotes lacking quantitative measures and controls. Lilly's writing differs from that of Herman Melville and Jules Verne not just in its more modest literary merit but more basically in its humorless and quite unjustified claim to be a scientific report.

With different quotations where I have left ellipses, this could be from an article by the SSG attacking Wilson.

> far off the correct path; it is better to have too much information than too little, especially when a discipline has only weakly defined its questions.

In the literal sense of this and similar passages, there is nothing but praise. But Wilson implies that an unacknowledged bias underlies the whole discipline, and even the praise at the end of the quoted passage places this discipline, which is at least several generations old, as still rather undeveloped and immersed in detail.

The responses to such descriptions by comparative psychologists like Frank Beach, Jay Rosenblatt, and Ethel Tobach may serve as typical of the responses of angry population geneticists, ethologists, and anthropologists. Although they make a number of fundamental criticisms, focusing on the failure to distinguish between levels of causation,[47] they do not establish universal scientific standards, and show how Wilson deviates from them, but instead defend the questions and practices of comparative psychology. Beach, for instance, does this by using loaded language and heavy irony to identify science with his discipline and nonscience with sociobiology (p. 133).

> The model-building sociobiologist may be unconcerned by a primatologists's criticisms to the effect that some reports of infanticide are of dubious reliability or that at most infanticide can be considered a rare form of behavior associated with abnormal ecological conditions. Such complaints may be seen as irrelevant by a theorist who is concerned neither with understanding the behavior of langurs as a species nor with analyzing the proximate causation of a particular behavioral incident. This attitude obfuscates effective communication with the comparative psychologist whose principal goals are to describe, measure, and compare analogous behavior patterns in different species and to analyze behavior in terms of its motivational and mediational components. To such an individual, the sociobiologist may fit in the category described by B. F. Skinner (1938, p. 44) as "men whose curiosity about nature is less than their curiosity about the accuracy of their guesses."

This passage does not show that sociobiology is not a science; it only shows that it is not good comparative psychology. Comparing this passage to Wilson's methodological parodies, we might think that

47. Professor Wilson argues that he has responded effectively to the criticisms of comparative psychologists, especially in his more recent work.

philosophy of science does not seem to provide the sciences with agreed-on methods, but only with a generally shared vocabulary of polemical abuse. Still, the fact that both sides use this vocabulary implies that they believe the controversy about human nature is to be settled on scientific grounds.

Science versus Ideology

One way in which Marxists have responded to the many texts that use sociobiology as scientific support for capitalism, sexism, or racism is to make a distinction between real science and ideology. This line of rhetoric, too, is a mirror image of that used by Wilson, when he accuses his critics of putting political considerations ahead of science. Wilson is, of course, using *ideological* in a sense significantly different from the various senses the term has in Marxist thought. Rose, Kamin, and Lewontin take their definition from Marx's *The German Ideology,* in which ideology is "the ideas of the ruling class" that "in every epoch are the ruling ideas"; ideology then is to be traced outward to the relations of production.[48] Wilson uses the term in the more popular sense, to describe the taint of politics in what is thought to be the nonpolitical; ideology then is to be traced inward to personal prejudices. Despite this basic difference, both sides use the the opposition of science to ideology to explain why other scientists could have other ideas. But both the explanation which sees ideology as the reflection of the interests of the ruling class and that which sees it as the dark irrational receding before the progress of science lead to some rhetorical problems. Both assume that there are agreed-on methods that allow us to tell real science from ideology. Thus neither questions the absolute authority of real science in the social arena. And neither narrative can account for the scientific beliefs the authors hold themselves; neither narrative can account for "real" science.

Wilson says in his response to the SSG's letter that it "is an openly partisan attack on what the signers mistakenly conclude to be a political message in the book."[49] As we have seen, all parties have rhetorical reasons to present disagreements as misinterpretations of an unambiguous text. In his more detailed response, "Academic Vigilantism," Wilson says he will account for the SSG's misinterpretation. "How is it possible for the Science for the People Group to misrepresent so consistently the content of a book, in contrast to all

48. *Not in Our Genes,* p. 4n.

49. Wilson, "For Sociobiology," p. 265.

the many reviewers among their scientific colleagues?" He starts with demographic terms, "the size and composition of the group," and one might think we are to get a sociological explanation of the opposition in terms of the population biology of academic groups, leading from certain environmental factors in Boston-area universities to demographic shifts in university faculties to political behavior. But the fact that he refers to the "political significance of sociobiology" in the title does not mean he is going to do a political analysis of his own science; he is referring to the significance conferred on the science by the activities of its opponents.

Wilson, like his critics, sees human affairs as governed by forces and relations that are hidden from them by a veil of unreality. In Wilson's case, the unreality is not capitalist ideology, but irrational impulses held over from earlier stages of the evolution of culture, especially from the adaptations of paleolithic hunter-gatherer societies. "Value systems are probably influenced, again to an unknown extent, by emotional responses programmed in the limbic system of the brain." It is part of the progress of science to tear away such illusions, letting us see these values as they really are, in terms of our present environment. The views of his political critics must be treated separately from those of his scientific critics, for he sees the political critics—just as they see him—as starting with nonscientific goals that lead to a commitment to one scientific theory.

> The belief-system they promote is clear-cut and rigid. They postulate that human beings need only decide on the kind of society they wish, and find the way to bring it into being. Such a vision can be justified if human social behavior proves to be infinitely malleable.

Marxism, then, is cast in the role the church has in nineteenth-century debates; he describes it in terms such as *belief system* and *vision:*

> When the attacks on sociobiology came from Science for the People, the leading radical left group within American science, I was unprepared for a largely ideological argument. It is now clear that I was tampering with something fundamental: mythology. Evolutionary theory applied to social systems is an extension of the great Western traditions of scientific materialism. As such, it threatens to trans-

> form into testable hypotheses the assumptions about human nature made by some Marxist philosophers.[50]

In his narrative, then, the persistent progress of science is opposed by both religion and Marxism, and both sources of opposition represent prescientific forms of thought that are defended by powerful interests, and that appeal to deep emotional (and therefore neurological) responses. (Thus, as Albury and others point out, defenders of Wilson repeatedly invoke such figures as Galileo and Darwin).

I noted in chapter 1 that the Strong Programme in the sociology of scientific knowledge requires "symmetrical" explanations that can account for "true" as well as "false" beliefs. Wilson, like the scientists studied by Nigel Gilbert and Michael Mulkay in *Opening Pandora's Box,* has two different kinds of explanation. The ideological beliefs of his opponents are to be attributed to social causes: their politics and their careers.

> In retrospect there appear to be two levels of meaning to their protestations. The outer meaning is the literal argument they gave, that "genetic determinism" of any kind will inevitably be used to justify reactionary political doctrines, racism, sexism, aggression, and other undesirable social responses associated with acceptance of the status quo. The deeper meaning, in my opinion, was the challenge they sensed to their own authority as natural scientists devoted to the study of social problems. ("Foreword," p. xiii)

But Wilson presents his own scientific beliefs as the outcome of processes entirely internal to science: the synthesis of population biology and behavioral ecology. In this account, it would be irrelevant to ask about Wilson's political views, or about Stephen Jay Gould's paleontology.

An account of sociobiology as ideological is given by the SSG:

> Determinist theories all describe a particular model of society which corresponds to the socioeconomic prejudices of the writer. It is then asserted that this pattern has arisen out of human biology and that human social arrangements are either unchangeable or if altered will demand continued conscious social control because these changed conditions will be "unnatural." Moreover, such determin-

50. On value systems, see Wilson, "Academic Vigilantism," quotations from pp. 297, 300, and 292, respectively. On Marxism, see Wilson, "Introduction," p. 2.

> ism provides a direct justification for the status quo as "natural," although some determinists disassociate themselves from the consequences of their arguments.

The "socioeconomic prejudices" of the writer are the motivating force behind the theory. Though such a theory can have no evidence to support it, the SSG argue, it is "seized upon and widely entertained, not so much for its alleged correspondence with reality as for its obvious political value." This is what Joe Crocker calls a "conspiracy theory" of sociobiology. The narrative is oddly parallel to the narrative of evolutionary adaptation in which environmental changes (here socioeconomic structures) lead to demographic changes (here ideological structures) which lead to certain behaviors (here historical events). Steven Rose says we must ask of the theory, "Who benefits?" But this is just what a sociobiologist asks of an instance of animal behavior.[51]

Sometimes texts by SSG members seem to put forward a view of the relation between science and ideology that sees all science as inherently political, but this view, if it is there, soon dissolves. So they say:

> Our central point is that sociobiology—like all science—proceeds in a social context; "pure objectivity" is as much a myth for sociobiologists as for science reporters. All attitudes towards sociobiology—ours as much as any—reflect certain political preoccupations which need to be made explicit.

This starts off by seeming, in the phrases set off by dashes, to admit a symmetry of interpretation in which the views of both sociobiologists and their critics would emerge from "social context." But the passage goes on to explain that social context affects attitudes towards ideas; it apparently does not construct the ideas themselves. And then it isolates sociobiology on a scale of the sciences as being further from facts and closer to human concerns: "The weaker the constraint of fact, the closer the subject to immediate human concern, the greater the influence of these preoccupations." Finally, the Marxist analysis applies only to sociobiology, not to all science. "What we have argued, and continue to assert, is that sociobiological ideas do not arise in a social vacuum but rather reflect the dominant interests and attitudes of the classes to which the authors belong."

51. SSG, "Sociobiology," pp. 280, 281; Rose, "It's Only Human Nature," p. 162.

When the SSG attack Wilson, they point out that he can give no account of how one gets from one level to another, from one stage of the theoretical narrative to the next—for instance, he does not actually locate specific genes for specific behaviors and show how they are selected and how they operate in the animal's development. But in the same way, they do not show, and do not need to show, exactly how Wilson's ideas, or anyone's ideas, actually arise from the forces and relations of production, or how Wilson's statement of these ideas causes events that would not otherwise have happened. Instead they argue, just as they say Wilson does, from analogy between two systems.[52]

The SSG have an explanation for why the ruling classes have the ruling ideas, and for how they are reproduced in books like *Sociobiology*. But no account is given in the SSG's text of "those of us who would change the way things are," of what interests and attitudes they reflect, or how they came into being in this socioeconomic environment. Another and more fundamental problem this view of ideology poses for Marxists is that it leaves intact the authority of science (that is, "good science" as opposed to the "bad science" of sociobiology) as something outside and opposed to ideology.[53] Rose,

52. Alper et al., "Implications of Sociobiology," quotations on pp. 334, 336, respectively.

Rose, Kamin, and Lewontin trace this form of argument to Boris Hessen's analysis of Newton in the 1931 collection *Science at the Crossroads* (see Werskey *The Visible College*, for an historical account), an analysis that has been enormously influential in later attempts to show the relation between science and society. Hessen's argument implies that if one set of concepts, such as Newtonian physics, can be seen to be structured like another, such as capitalist exchange, then it can be assumed that the social system caused the concepts in the scientific system, even though the details of cause and effect would be far too complicated to trace. Similarly, Wilson's critics assume that if they can show that his text incorporates terms and concepts from capitalist economics, which it does, then it both arises from and contributes to capitalism.

53. One critique of sociobiology that does not rely on the distinction between good science and bad science is an article in the *Radical Science Journal* by Joe Crocker, "Sociobiology: The Capitalist Synthesis." Crocker supports the conclusions of Steven Rose's critique, but rejects Rose's view of science: "Because ideology and truth are mutually exclusive in Rose's philosophy of science, it would be disastrous for his politics if sociobiology were in any sense true. From the start, he is obliged to dismiss it as false. In this article, on the other hand, it is insisted that all science ("good" or "bad") is incorrigibly ideological. Sociobiology is ideological precisely because its practitioners aspire to be good scientists in the tradition of Newton and Galileo, seeking to deduce universal social laws from individual behavior" (p. 61).

The problem with this approach for many scientists is that Crocker would see ideology, not only in the argument of sociobiology, but in such basic processes as quantification. "Scientific categories," he says, "are *constituted* by social relations such

Kamin, and Lewontin argue that a science that draws its terms and concepts from dialectical materialism does not arise from capitalism, and will not lend scientific authority to the maintenance of the capitalist system.[54] But such an approach just shifts the scientific metaphor from optimization to dialectic. One alternative to such tactics, I would argue, is to promote skepticism about the production of all knowledge and about its authority in political discourse. In this view, we do not undermine the political authority of sociobiology by saying that Wilson is just telling stories. Wilson is telling a story very well, and the only effective answer to a story, as I think his critics show, is to tell another story.

The Polar Bear and the Whale

I began this chapter by noting the surprising persuasive power that *Sociobiology* has even for someone like me who came to it forearmed with criticisms, and who still does not agree with it. One of the anomalies that the analyst of rhetoric finds in the controversy that followed the book is that neither side seems particularly interested in persuading anyone who does not already agree with them. Indeed, their texts, like the first SSG letter, or like the aggressive passages at the beginning and end of *Sociobiology*, would tend to alienate anyone coming to them with even slightly different views. In this sense the debate never seems to get anywhere, and yet it goes on and on. I have noted that the methodological arguments of the two groups of experts do not lead to a resolution in which one side can clearly claim the authority of science, because there is no agreed-on method, but only a shared rhetoric of praise and abuse, and because in practice each scientific discipline takes its questions and practices as the standard.

Two sociological analyses of controversies have suggested functions for such debates, both focusing on the way the participants appeal to an audience outside science. Yaron Ezrahi notes a sort of rhetoric similar to that we have seen in the sociobiology controversy in his analysis of the controversy over alleged links between race and lower I.Q. test scores.

---

that the world is comprehended in their image" (p. 64). Such a position requres that one give up a lot; as the last sentence of the passage suggests, Crocker would call physics into question as well as sociobiology. This does not lead Crocker to the acceptance of sociobiology or the rejection of all scientific knowledge, but it does lead to the rejection of any scientific claim to objective knowledge of a world that transcends social processes.

54. *Not in Our Genes*, pp. 265–90.

> My first contention is that the strategies of the principal participants in the controversy prove that the controversy was not focused on the effort of settling differences of scientific opinion by the application of scientific norms of discourse and proof, but was primarily a contest between rival scientific groups and their respective supporters over which definitions of fact and reality would be certified by the collective authority of science as valid premises of public policy or social interaction. The issue was not one of defining social fact but of establishing as facts in the social context certain propositions.[55]

I would not see such rhetoric in either the race/I.Q. controversy or the sociobiology controversy as a sign of deviance from scientific norms under social pressure, for as I have argued in earlier chapters, I see the "scientific norms of discourse" as another kind of rhetoric. But Ezrahi's suggestion that such debates are about rival claims to authority can help us understand the odd construction of these texts. They must address the public audience while seeming not to, for in the competition among scientists for public authority, to acknowledge that one wants political authority is to disqualify oneself as a scientist.

Ezrahi's analysis can help us understand the odd tone of texts that do not seem intended to persuade their nominal scientific audience. Ullica Segerstrale, in an excellent analysis of interviews with participants in the sociobiology debate, can help us understand the persistence of the controversy over the last ten years. Segerstrale argues that both sides have an interest in keeping the controversy going.

> Once the sociobiology controversy began, strategic interests came into play on both sides. As the debate developed, it was in neither party's interest to straighten out misunderstandings—instead the point became to develop one's own position while dismissing the opponent's one as "extrascientifically motivated. This way Lewontin let Wilson graduate to a leader first of the "adaptationist" and later of the "reductionist" program, while Wilson chose to retain Lewontin as a useful strawman for *tabula rasa* "Marxist" environmentalism.

55. Yaron Ezrahi, "The Authority of Science in Politics" in *Science and Values*, ed. A. Thackrey and E. Mendelsohn (New York: Humanities Press, 1974), p. 232.

In quoting a comment on the race/I.Q. detate, I do not mean to imply that it is necessarily the same issue as the sociobiology debate. As I have noted, such a claim is part of the political genealogy the SSG tries to draw for Wilson. For a Marxist analysis of the I.Q. controversy, see Les Levidow, "IQ as Ideological Reality," in *Radical Science*, ed. Les Levidow (London: Free Association Books, 1986).

Segerstrale's analysis helps us understand why neither side makes the sort of moves towards consensus one might expect in a controversy within a scientific discipline. Though I find Segerstrale's analysis of Wilson's and Lewontin's views very useful, I do not think that the origins of the controversy can be traced to the personal careers and goals of two scientists.[56] Segerstrale draws on extensive (and fascinating) interview material, and that may be why in that analysis the basic causes seem to be on the level of the strategies of individual scientists, as in the quotation given above. My study has drawn only on some of the published texts in the controversy, and in them one gets no sense of authors in control of their own rhetoric, nor of the wide divergence of strategies. The characters of Wilson and Lewontin, central to Segerstrale's analysis, seem to me to be creations of the controversy, not the creators of it. The two sides do not produce two different kinds of texts, in accordance with their radically different views about science and society. Instead, their rhetorical tactics seem to reflect each other, and their phrases and arguments echo back and forth, as if each text was made out of bits and pieces of the preceding text, reinterpreted to read ironically.

A number of studies cited in chapter 4 have shown how a limited controversy within a scientific discipline (geology, high-energy physics, biochemistry) tends to be resolved in the course of repeated interpretation such that the whole controversy, and the losing side, are forgotten, and all the remains is a fact. The participants in the sociobiology debate act as if it too can be resolved without leaving a trace, as soon as *their* facts are accepted by the public as *the* facts. Indeed, the tendency in more recent writings has been to act as if this has already happened, so that critics tend to act as if sociobiology were publicly discredited, whereas sociobiologists tend to act as if the political criticism were a thing of the past (neither of which views is supported by a survey of articles published in the last few years). The story of the sociobiology controversy looks more like that of the controversy over *Cnemidophorus* we saw in chapter 4, in which the two

56. Segerstrale, "Colleagues in Conflict," p. 79. The conflict between Wilson and Lewontin is a good story, and the story was used in some journalistic accounts of the controversy (see, for instance, Colin Campbell, "Anatomy of a Fierce Academic Feud," in the *New York Times*, which is based on Segerstrale's article). The SSG is right to say this focus trivializes the debate, which is about more than personalities. Despite the title of Segerstrale's article, it is not just about personalities, and it says at the outset, "the sociobiology controversy would be misconstrued if it were seen as merely 'an in-house quarrel between Harvard professors,' whether politically motivated or not" (p. 54).

sides continue their own lines of work, without addressing the other. Such nonresolution is even more likely in the sociobiology debate. In a controversy between members of different disciplines, a controversy about the origins and future of human society, a controversy between what Segerstrale calls "scientific-cum-moral agendas," no facts are likely to be agreed on.

In "From the History of an Infantile Neurosis," Freud comments on his ongoing controversy with Jung and Adler, and their failure to agree on what he considered basic postulates: "The whale and the polar bear, it has been said, cannot wage war on each other, for since each is confined to his own element they cannot meet." There is no reason why the sociobiology controversy should not continue indefinitely.[57]

57. Professor Wilson disagrees with my conclusion that the controversy can go on indefinitely. But then, when Freud used the metaphor of the polar bear and the whale, he did not mean that his controversy with Jung and Adler could go on forever; he was pointing out that it was not worth arguing with them when they would not accept his postulates, and was implying that his argument would triumph.

Conclusion
Reading Biology

Any treatment of science focusing on its discourse is likely to be seen at first, especially by nonscientists, as an attack on its cultural and political authority as the underwriter of reliable knowledge of nature. Some welcome this attack as a way of turning the tables on the pretensions of now-dominant disciplines, whereas others see it as a dangerous opening that allows other kinds of religious or political belief to take the place of objective science. For instance, several readers' comments on my chapter 6 suggested that because I would not see the rhetoric of *Sociobiology* as essentially different from that of its critics, I was ceding ground to right-wing, sexist, or racist views that have taken sociobiological texts as their scientific authority. These comments, coming as they do from those who share many of my assumptions, demand a response. In the first chapter I argued on methodological grounds for a focus on scientific texts—they can tell us things about science and about texts that we would not learn by other approaches. Here I respond to a different question, a tactical question: can this approach to science be a basis for a good political strategy for the social sciences and for the public, or does it undermine effective social action? I shall also give some idea how I would expect this tactic to work in practice by outlining some of the strategies I would like to see nonbiologists employ in reading biology.

I do not see the analysis of scientific texts as a project of debunking science. Those who have seen these studies as critical are, I think, comparing the view I present to a particular and very restricted realist and empiricist view of science that leaves no room for texts at all. For those holding the realist view, any mention of rhetoric would be debunking: grant proposals, they would argue, are awarded solely on an abstract and quantifiable measure of scientific merit, and they would point to the agency's numerical scores and funding cutoff, and the quantitative approaches of some science policy, as supporting their view. For these readers, my argument in chapter 2, that grant proposals involve a rhetoric of self-presentation and of placing oneself in relation to the field, and that the writing of them involves and reflects a complex process of negotiation with a particular audience—the agency's panel—could be seen as an attack on scientific objectivity. If one believes that scientific articles merely carry information, and that they are either judged on the truth and significance of that information, or misjudged on the basis of personal or political bias, then my argument in chapter 3—that the form of scientific claims is negotiated in the processes of writing and refereeing—could be seen as an attack on the reliability of scientific information. If one believes that science is an unmediated encounter with nature that leaves no room for controversy, or that controversies are resolved by the discovery of facts and the application of logic, my argument in chapter 4 that controversy involves the construction of stories and the ironic reinterpretation of these stories would seem to reduce scientific argument to trivial word games.

The implications of my last two chapters, on popularizations, are somewhat different, for they deal with the texts on which nonscientists depend for their view of science. In the most restricted realist and empiricist view, scientific facts are objective statements about the world and remain everywhere exactly the same whether they are stated in the *Proceedings of the National Academy of Sciences* or in *Scientific American* or in *Time*. In a public controversy like that over sociobiology, this view of science would suggest that the public need only find out the facts, or, in a case where there is conflict, find out which of the competing disciplines adheres most closely to the methods of science prescribed by philosophers. But if, as I have suggested, different audiences get different narratives, and different narratives carry different views of the work of science, those who take a realist and empiricist view of science can only ask which narrative is correct. If one starts with the assumption that science discovers objective facts solely through empirical methods, then my attention to texts in each study here will seem to be either an attack or a distraction.

The view of science that allows one to be surprised by my findings about scientific texts is indeed a limited one, and it is not the view of science held by most scientists, at least by those I have encountered. In contrast to groups of nonscientists to whom I've spoken, scientists have neither been surprised nor felt threatened by my comments on scientific rhetoric. On the very rare occasions when the subjects of these studies have asked for a change or omission, it was always because I had left room for the implication that they or someone else was guilty of fabrication, incompetence, or bad management, or where they could be seen as criticizing or mocking other scientists.

Such implications could be seen as unscientific behavior, but my discussion of their rhetoric never seems to have been taken as an attack on them as scientists. To take one example, many people have seen the first part of chapter 6, where I discuss the construction of E. O. Wilson's *Sociobiology*, as a criticism of the book. But Wilson himself does not seem to have seen it in this way.

Although I do not see this book as a debunking project, I do see the analysis of scientific texts as a way of promoting change. Specifically, I would like nonscientists to read more science, to read it more critically, to read it with an awareness of the social processes that produce it, and to question the authority with which science is sometimes presented in cultural and political contexts. These might seem uncontentious and humane goals. But analyses like mine have provoked fierce criticisms from other analysts concerned with science and society, especially from those on the left with whom I would, in general, like to agree. Some on the left have seen the attention to academic science as irrelevant, the focus on texts as trivializing, and the commitment to relativism as dangerously idealist. For Marxists, this is a tactical issue; we will not be able to determine which view is correct, but we can see how the views affect the actions of people.

One tactical question that might be raised is whether one needs to study biology at all. Most critical discourse analysis has focused on such obviously political material as newspaper reports, or political speeches and talk about political issues.[1] But Steven Rose, Stephen Jay Gould, J. B. S. Haldane, and others have pointed out the political importance of biology. And I would argue that a treatment of biology needs to deal with the practices of the discipline in detail; the important points cannot be isolated as "concepts" or "themes" accessible to those outside the discipline. So an analysis of biology articles has, potentially, as much political interest as an analysis of the Pentagon Papers or the Watergate tapes.

Potter and Wetherell raise another tactical question about textual study:

> People sometimes assume discourse analysis denies the existence of a world "out there." "Why this concentration on language," they ask, "when people out there are giving birth, making money, and being murdered by oppressive regimes? Why don't you study these

1. Diane Macdonell, *Theories of Discourse: An Introduction* (Oxford: Basil Blackwell, 1986); Gill Seidel, "Political Discourse Analysis." *Handbook of Discourse Analysis*, Vol. 4, ed. Teun van Dijk (London: Academic Press, 1985), pp. 43–60.

> real processes and not just language which is a second-hand superficial medium?"

Their answer to this question (which admittedly is put in a rather overstated form) is the argument that there is no dichotomy "between 'real' events and linguistic representations of those events."[2] Potter's and Wetherell's own work, and that of some other discourse analysts focuses on the discourse of racism, and is hardly likely to be dismissed as irrelevant to the world "out there." The texts I study are not obviously political, but they are central to the processes of constructing facts, methods, and authority, in a field that is central to our view of ourselves and of society. While radicals within biology have tried to separate good biology from bad, I am arguing that nonbiologists need to know about the textual processes by which any biology is constructed.[3]

Some analysts of biology would object on political grounds to the relativism of my approach. Hilary and Steven Rose identify the Strong Programme in the sociology of science (the basis for my analysis) with other, more obviously reactionary threats to the "radical science" Rose and Rose had welcomed in several influential publications earlier in the 1970s.

> It is this philosophical relativism which has moved from being a critique of other knowledges to an auto-critique of one's knowledge and on towards an escalating reflexivity. It is a hyper-reflexivity spoken of as the "disembodied dialectic" which, both within the sociology of scientific knowledge and within the radical movement, threatens to consume not only ideology but science itself. The certainties of the Althusserian disinction between scientific knowledge and ideology are to be obliterated, dissolved into their social determinations and a belief in the equality of discourses.

Apart from the rather peculiar jargon attributed to relativists, this could be taken as a fair summary of the relativist project and the response to Althusser. But by the end of the paragraph, the position they are criticizing leads, seemingly inevitably, to trivia. "To be cool, to be aware that we are playing in nothing more than a series of more

2. Potter and Wetherell, *Discourse and Social Psychology*, pp. 180, 181. Steve Woolgar presents an argument in more detail in "Discourse and Praxis."

3. On racist discourse, see the Spring 1988 issue of *Text*, edited by Teun van Dijk. On good and bad biology, see, for example, Rose et al., *Not in Our Genes*.

or less elaborate games, constitutes the new authenticity. The politics of subjectivism replace the pursuit of the rational society." In this view, there is no position between the acceptance of the authority of science on one hand and intellectual dandyism on the other. Any questioning of the special position of scientific knowledge "disarms radical scientists."

> For instance, when the ideologues of scientific racism, such as Jensen and Eysenck "work on knowledge," is what they produce new knowledge, and if not, what is it? If it is fetishized consciousness, as RSJ [the *Radical Science Journal*] argues, there are no rational grounds for opposing it and the opposition to scientific racism must be seen exclusively in personal and moral terms. If we adopt the position of what is called 'the strong programme' in the sociology of knowledge we must presumably regard all these cultural products as new knowledges.[4]

This sounds as though the RSJ and the Strong Programme would tend to support scientific racism. But that conclusion comes because the passage makes an odd hybrid of the Strong Programme view of knowledge as beliefs about reality and another view in which objective knowledge of reality is the only basis for action. The combination of a skepticism about scientific knowledge and a belief in the absolute authority of some given body of knowledge would indeed be disastrous—the creationists in the United States are an example. But the Strong Programme is not asking what is really the case, it is asking why people believe what they believe.[5] The practical effect of the acceptance of their argument would not be an acceptance of racism, but a skepticism about all claims to scientific authority.

The political strategy of discourse analysis is based on the assumption that this skepticism about scientific authority is a good thing. For instance, some of the current local and national political issues that involve biological expertise include the government directives to farmers after Chernobyl, the possible statistical evidence for increased leukemia around our local nuclear power plant, the use of recombinant DNA technology in agriculture and pharmaceuticals, and the methods

4. Hilary Rose and Steven Rose, "Radical Science and Its Enemies," in *The Socialist Register, 1979*, ed. R. Miliband and John Saville (London: Merlin, 1979), pp. 317–35; quotations pp. 324, 326, respectively (footnotes omitted).

5. A Marxist response to Rose and Rose along these lines is the article by Joe Crocker, "Sociobiology: The Capitalist Synthesis."

used to quantify "quality of life" in the British National Health Service management.[6] In any of these cases, a critical approach to expertise is politically useful. An analysis that would translate the scientific issues into more recognizable political terms such as those of class struggle, state apparatuses, and forces and relations of production might have some use in the case of, say, Chernobyl, but what would be most useful to, say, Cumbrian sheep farmers, would be an understanding of the function and limits of expertise, and of its relation to their own practical expertise in such matters as grazing, weather, and sheep behavior. Such questions arise, not just with obviously controversial issues such as radiation risks, but in nearly every scientific and technological story in the news.[7] Of course, skepticism alone is not a political program, and Rose and Rose are right to point out the dangers in this direction. But it is a reasonable part of a political program. And it is particularly important when we realize that the political program is itself a part of discourse, that it needs to construct narratives and reinterpret the narratives of others if it is to be persuasive. One consequence I would like to see come out of this book, and out of many of the studies in the sociology of scientific knowledge, is a change in reading habits that would make us more active, critical readers of the scientific discourse that enters into our lives.

6. On Cumbrian sheep farmers' responses to experts, see Brian Wynne, Peter Williams, and Jean Williams, "Cumbrian Hill-Farmers' Views of Scientific Advice," *Evidence to the . . . Select Committee on Agriculture: The Chernobyl Disaster . . .* (Lancaster: Center for Science Studies and Science Policy, University of Lancaster, 1988); see also Brian Wynne, "Establishing the Rule of Laws: Constructing Expert Authority," in *Expert Evidence: Interpreting Science in the Law* (London: Routledge & Kegan Paul, 1988), and "Unruly Technology: Practical Rules, Impractical Discourses and Public Understanding," *Social Studies of Science* 18 (1988): 147–67; and, on health economics, Malcolm Ashmore, Michael Mulkay, and Trevor Pinch, *Health and Efficiency* (Milton Keynes: Open University Press, 1989).

7. The following news report (July 28, 1987, "You and Yours," BBC Radio 4) illustrates a problem readers may have with any study trying to show the social construction of scientific knowledge. A court had ruled that a defendant could be convicted of drunk driving, even if his alcohol level was below the legal limit when tested four hours after the accident, because a "scientific" method of "back calculation" could show that his alcohol level had been excessive earlier when he was driving. But a representative of the Police Surgeons Association had expressed doubts about the use of this method in a judicial context, saying there was disagreement within his organization about whether any individual case might have a radically different rate of absorption from the average of many cases that was used as a basis for extrapolation. So, the reporter immediately responded, the method is not *really* scientific. This is the kind of Catch-22 that keeps us from ever seeing the social in the scientific. At first, the fact that the method is "scientific"—without any qualification about what this might mean—puts it beyond any legal or political decision-making. But then, as social factors become apparent, the method is no longer seen as scientific.

I shall try to illustrate some strategies for critical reading with two texts as examples. One is a passage from a book on childrearing that claims that its conclusions are based on scientific research, *Girls and Boys: The Limits of Non-Sexist Childrearing,* by Sara Stein. It may seem too easy to be critical about such a book, but it can stand for a number of other, more sophisticated uses of scientific authority in popular texts. The passage is from a chapter titled, "Ten Million Years of Sexism."

> [U]ncovering the origins of sexism will shed light on the puzzle we still haven't pieced together. Park bench impressions and laboratory measurements concur; there are sex differences. But what is the point of them? Or are they, as feminists claim, beside the point?
>
> If our peculiarly human forms of sexism were invented, we would be surprised to find them in a close relative of man. But if they evolved, we should not be surprised at all. Our closest living relative is the chimpanzee. And chimpanzees are sexist.
>
> Female chimps stay close to home. Males spend their day far afield. Females cluster with one or two best friends, their daily routine of gathering staple foods interrupted only to nurse a baby, break up a squabble, or scold a straying toddler. They love to fish for termites by poking sticks into the mound and nibbling off the soft, plump insect that cling to it (p. 31).

My other text is also about the social habits of primates; it is a table reproduced in Wilson's *Sociobiology* from an article Crook and Gartlan published in *Nature* in 1966 (see p. 254). Its validity as a set of categories is not my concern here. Nor is it my purpose, in choosing Stein's book from all the many popular texts that draw on sociobiology, to criticize either the anthropomorphism or the pseudoevolutionary argument of such popularizations.[8] What interests me here is how nonbiologists might approach such popular and specialized texts. Reviewing the studies in this book, I think of five strategies I would apply to both texts.

8. Similar arguments are analyzed in Rose et al., *Not in Our Genes;* in Dialectics of Biology Group (S. Rose, general editor), *Against Biological Determinism* (London: Allison & Busby, 1982), and in Lynda Burke and Jonathan Silverton, eds., *More Than the Parts: Biology and Politics* (London: Pluto Press, 1984). For that matter, Wilson's chapter on primates in *Sociobiology* gives enough explanation of the difficulties of reasoning from primates to man to make one sceptical about this particular argument, even though Stein presents sociobiology as the authority for her argument.

Table 1. ADAPTIVE GRADES OF PRIMATES (SEE TEXT)

Species, ecological and behavioural characteristics	Grade I	Grade II	Grade III	Grade IV	Grade V
Species	*Microcebus* sp. *Chierogaleus* sp. *Phaner* sp. *Daubentonia* sp. *Lepilemur* *Galago* *Aotus trivirgatus*	*Hapelemur griseus* *Indri* *Propithecus* sp. *Avahi* *Lemur* sp. *Callicebus moloch* *Hylobates* sp.	*Lemur macaca* *Alouatta palliata* *Saimiri sciureus* *Colobus* sp. *Cercopithecus ascanius* *Gorilla*	*Macaca mulatta*, etc. *Presbytis entellus* *Cercopithecus aethiops* *Papio cynocephalus* *Pan satyrus*	*Erythrocebus patas* *Papio hamadryas* *Theropithecus gelada*
Habitat	Forest	Forest	Forest–Forest fringe	Forest fringe, tree savannah	Grassland or arid savannah
Diet	Mostly insects	Fruit or leaves	Fruit or fruit and leaves. Stems, etc.	Vegetarian-omnivore Occasionally carnivorous in *Papio* and *Pan*	Vegetarian-omnivore *P. hamadryas* occasionally also carnivorous
Diurnal activity	Nocturnal	Crepuscular or diurnal	Diurnal	Diurnal	Diurnal
Size of groups	Usually solitary	Very small groups	Small to occasionally large parties	Medium to large groups. *Pan* groups inconstant in size	Medium to large groups, variable size in *T. gelada* and probably *P. hamadryas*
Reproductive units	Pairs where known	Small family parties based on single male	Multi-male groups	Multi-male groups	One-male groups
Male motility between groups	—	Probably slight	Yes—where known	Yes in *M. fuscata* and *C. aethiops*, otherwise not observed	Not observed
Sex dimorphism and social role differentiation	Slight	Slight	Slight—Size and behavioural dimorphism marked in *Gorilla*. Colour contrasts in *Lemur*	Marked dimorphism and role differentiation in *Papio* and *Macaca*	Marked dimorphism. Social role differentiation
Population dispersion	Limited information suggests territories	Territories with display, marking, etc.	Territories known in *Aloutta*, *Lemur*. Home ranges in *Gorilla* with some group avoidance probable	Territories with display in *C. aethiops*. Home ranges with avoidance or group combat in others. Extensive group mixing in *Pan*	Home ranges in *E. patas*. *P. hamadryas* and *T. gelada* show much congregation in feeding and sleeping. *T. gelada* in poor feeding conditions shows group dispersal

Crook and Gartlan's table of the Adaptive Grades of Primates from "The Evolution of Primate Societies," by J. H. Crook and J. S. Gartlan, *Nature*, Vol. 200, p. 1200. Reprinted by permission.

1. Look for the rhetorical. The rhetorical element is clear enough in Sara Stein's book; as the title suggests, she wants to convince parents that the gender roles currently defined in our society are natural and inevitable. She steps back from public debate on such matters to the authority of science, and, in a series of moves, comes to the conclusion that whatever traits are shared by humans and chimpanzees are unchangeable in man. Then she gives as the scientific facts a highly anthropomorphized account of chimpanzee behavior. As the blurb on the back of the book says, it is "a rare and reassuring blend of park-bench common sense and wide-ranging specialist research." Although the blend may be reassuring, it is hardly rare; it is a staple of popular scientific comment on human behavior.

In this book I have argued that we do not escape the rhetorical by going back and back to more and more specialized scientific works—to Wilson, and then to researchers like Crook and Gartlan, and even to such a seemingly arhetorical text as a table. For this writing, too, is rhetorical. Species are arranged so as to contrast certain features of their behavior and suggest relations of social organization to ecological facts. This table is not a way of presenting new data, as in, say, a table in an analytical chemistry article; instead it arranges information known from recent studies so as to suggest a new view of the evolution of primate behavior. So this table is not just a representation of what is known but an attempt to make other primatologists accept a claim about this knowledge. And this claim is likely to be contentious; for instance, when Wilson cites it he goes on to give an alternative table that he finds more persuasive. Each element of the table is also a potential matter for controversy. We should recall that Crews and his critics could produce tables of lizard behavior that suggested different interpretations of the relevance of the behavior to reproduction. Bloch, too, rearranged known data in a table that made a new and controversial claim. So such presentations of information are rhetorical, and never more so than when they seem simply to reproduce objective knowledge.

2. Reconstruct the social context. We have to remind ourselves of the social context of any scientific text because the form of scientific texts conceals the social—science covers its own tracks. So, for instance, Stein's appeal to scientific knowledge does not allow for the possibility of disagreements among primatologists, or for consideration of the traditional practices of ethology, or for questions about why the research was undertaken. But these omissions are not just matters of simplification or vulgarization of more complete scientific originals. Crook and Gartlan's table does not, by itself, suggest that there could be any controversy about such a scheme. It cites its sources but does not say how various studies were made comparable for the purposes of this exercise in comparison, and does not show how this information was gleaned from a number of other texts. These issues would come out only if there were a controversy about the claim suggested by the table. Similarly, a reader of the latest texts in the *Cnemidophorus* controversy (chapter 4) would think that people in the field surely agree, for in each article the opening summary of the literature and the closing remarks present a firm and growing consensus. And a reader of Bloch's or Crews's published articles (chapter 3), after all their changes, has no way of knowing that the author originally wanted to say something quite different. The very form of the articles and the manner of their publications work against any indication of the social context.

3. **Look for related texts.** One way of seeing this social context is by restoring each of the fragments I have quoted to the context of other texts. In the case of the popularization one would start by looking up Jane Goodall and other primatologists, looking up other books by Stein, looking up other uses of sociobiology to define gender. D. R. Crocker has done this adeptly for just the sort of research and popularization of primate behavior we see here, to show the differences in the anthropomorphization in popular and specialized texts. The critiques of sociobiology make other sorts of political connections. Bruno Latour and Shirley C. Strum have linked such accounts of the origins of society to an even broader survey of philosophical and historical texts.[9]

But Crook and Gartlan's text also emerges from other texts. The authors give us some help here; they say that the table draws on recent work in the field, but also, and more surprisingly, on a model developed in ornithology and on evidence from paleontology. At the end of the article we see that this set of categories can be used to evaluate the hypotheses of other primatologists. And when we see the table in Wilson's book, we see how it is incorporated—or not—into later research. A critical reading of the table would lead us into these various subdisciplines and their texts. The making of such a

9. D. R. Crocker, "Anthropomorphism: Bad Practice, Honest Prejudice"; Latour and Strum, "Human Social Origins."

table affiliates Crook with some approaches to the study of primates in evolutionary terms, rather than with others. This sort of reconstruction and critical reading begins with comparison. Although the reader of Crews's *PNAS* article would not see the possibility of alternative views of the same set of actions by the lizards, the reader of the whole *Cnemidophorus* file, with all the articles, would begin to see different methods, different research goals, different philosophies of science. The nonscientific reader who reads a biology article might do well to start by figuring out the rhetorical point of each of the citations, as I have tried to do with Crews's and Bloch's citations in the articles studied in chapter 2.

4. **Look for the source of authority.** The authority of the passage from Stein, and of the whole book, is based on the happy discovery that "Park bench impressions and laboratory measurements concur: there are sex differences." She presents two kinds of authority, the authority of objective science and the authority of subjective common sense. And the great thing, she says, is that the two agree. But that is not surprising, since her heavily anthropomorphized descriptions of chimpanzees impose conventional social language on them, to find that they are just like humans. (The same sort of turn is done in a *Business Week* article that notes that sociobiology confirms Adam Smith's economics, "A Genetic Defense of the Free Market." But this is hardly surprising, because, as Wilson says, sociobiology takes its model of optimization from economics in the first place.) Science, for Stein, must be true because it confirms what we already know. This kind of authority contrasts with the other indicator of scientific authority in Stein's book, the references to individual scientists as experts, with the names of their institutions. Sometimes it seems the scientists are right because they confirm common sense, sometimes because they are uniquely gifted in abstruse knowledge. We saw the same sort of tension in the *Time* report of the *Cnemidophorus* research in Crews's lab.

The authority of Crook and Gartlan's table depends more on a different sort of authority, a different sort of consensus. It is persuasive because it marshals the studies of many different researchers and makes them all work to one end, to one claim. And the claim in it becomes more fact-like as more and more researchers believe it, use it, base other statements and other work on it. This cumulativeness of scientific texts is what gave Crews and Bloch trouble with the articles in chapter 3. At first Bloch's claim didn't seem to relate to anything, didn't seem to have interesting work for other scientists in it. Crews's claim, on the other hand, threatened the work of other researchers; if he was right, they had to change their approach, and look at an aspect of behavior in relation to hormones that they hadn't considered before. The tight linking required in scientific discourse between one text and others is part of what makes such texts so forbidding to nonscientists who pull out one block in the middle of this pyramid of claims.

5. Look for any links between scientific language and everyday uses of language. Though scientific texts come out of an unusual social structure, and thus are different in some details from texts in other discourses, they are not doing something fundamentally different from other texts. We may stress too much the inaccessibility to nonspecialists of scientific writing. Science uses our language, and despite all attempts to purify it, it is still loaded with social and political implications. These implications are clear enough in Stein's passage. The whole passage, the whole chapter, depends on using the language of sexual stereotypes for descriptions of animal behavior. But as Crocker and others have argued, it is very difficult to eliminate anthropomorphism from behavioral studies. Similarly, the language of biology enters other discourses—such as that of childrearing. Our goal as critical readers should not be to purify the language either of everyday descriptions of behavior or of ethology, but to trace the movements back and forth between the discourses.[10]

If we are to track these textual transformations as critical readers, it is crucial that nonscientists not treat scientific texts as some sort of foreign language. If some of the passages I have analyzed in this book seem forbidding in their vocabulary and methods, this should not conceal the fact that they work just like other texts in English. We know, when we read newspaper articles on how to bring up one's baby, or when we read a letter from a solicitor, or even a poem, that we are stepping into areas of controversy, of rhetoric, of social conflicts. We do not read them as simply communicating data. I have argued in this book that the same is true of scientific texts; they must be put back in the social context from which they arise. As Frederic Jameson says, introducing his collection of readings of novels, "Inter-

10. Teri Walker, "Whose Discourse?" in *Knowledge and Reflexivity: New Frontiers in the Sociology of Knowledge*, ed. Steve Woolgar (London and Beverly Hills: Sage, 1988), pp. 55–79.

pretation is not an isolated act, but takes place within a Homeric battlefield, on which a host of interpretive options are either openly or implicitly in conflict."[11] The scientific articles that may seem to the nonscientists to be fixed and conventional formats, filled with the appropriate facts and jargon, need to be seen as the battlegrounds on which the terms of knowledge are being defined. That we do not see the armies of the other interpretative options—the losing views of phenomena—is only because in this battle, the losing army is immediately buried. We see only the shining armor of the facts that remain.

11. Fredric Jameson, *The Political Unconscious: Narrative as a Socially Symbolic Act* (Ithaca: Cornell University Press, 1981), p. 13.

Appendices

References

Index

Appendix 1: Texts of Proposal Summaries

Dr. Bloch's Abstract

Version 1

1. A search is being conducted for sequence homologies and for homologies of the reverse complementary sequences among tRNAs and rRNAs. 2. The results of these searches are being compiled in order to compare the distributions of the sequences among different regions of the molecules.

3. The purpose is to search for evidence of common origins of these classes of RNA.

4. A model is proposed for the evolutionary origin of the protein synthetic mechanism which predicts a common origin of the different classes of RNA. 5. The model is based on a synthesis of a multifunctional RNA through a series of alternating syntheses: elongation through looping back, replication via templating, then repetition of the this cycle [sic], starting with a primoridal tRNA with a simple anticodon region. 6. The result would be a molecule with extensive internal complementarity, dotted with codons and anticodons, capable of assuming configurations that would permit it to serve as message and structural RNA, and alternatively as gene.

7. The model predicts extensive homologies between the primordial tRNA and rRNAs. 8. Homologous sequences in present day tRNAs and rRNAs are being found.

9. An attempt is being made to sort out the relative importance of function and common descent as explanations for the homologies, by studies of the commonality of the homologies and their placements within and among the RNAs of the different classes.

Version 2

1. Ribosomal RNA is peppered with tracts that are homologous with regions found among the different transfer RNAs. 2. The matches are too frequent and extensive to be attributed to coincidence. 3. Their distributions and patterns suggest a common evolutionary origin for the two classes of molecules. 4. Function as an explanation for their existence

appears unlikely but cannot be ruled out. 5. Different domains have been conserved in different classes of organisms. 6. Our work will continue to identify examples of these homologies by searching for them among a variety of organisms. 7. The search was prompted by a model for the origin of a primitive multifunctional RNA molecule. 8. In the model, a short RNA with a codon or anticodon near the 3′ end undergoes successive rounds of elongation by self-priming (looping back) and self-templating, giving rise to an RNA in which codons are held in contiguous configurations by secondary folding. 9. The subsequent split of message, transfer, and ribosome functions is thought to follow acquisition of the cellular habit. 10. The model suggested the existence of homologies among present day t- and r-RNAs and this prediction is being realized. 11. The interpretation of the homologies is of importance. 12. A multidimensional test for evolutionary convergence has been designed and is being used to determine whether the homologies do indeed reflect common origin rather than function. 13. Filling out the rRNA map, through continued accumulation of homologies, should permit the reconstruction of a primordial RNA molecule.

Version 3

1. A large minority of tRNAs from all species of organisms studied have stretches whose base sequences are identical or nearly so to stretches found in rRNAs.

2. They are too frequent and too extensive to be attributed to coincidence.

3. Factors contributing to these matches might be shared functions at the RNA or DNA levels, or common origins. 4. The latter might be of recent derivation through recombination and transfection, or relics of ancient origin. 5. The matching sequences are distributed without discernable pattern among the molecules and among species. 6. Their frequent appearance, often unique to interspecies comparisons, indicates that they need not result from selection for interaction in a common cellular environment. 7. They are also thought to be conserved vestiges of ancient origin. 8. The occurrence of overlapping sets of homologies within species, and confirming overlays among species (homologies found in independent searches in different organisms that occupy equivalent positions on the rRNAs, and assign similar base sequences) suggest that their continued identification should permit the reconstruction of an RNA that is ancestral to both tRNAs and rRNAs. 9. Such a "synthesis" should help to provide an understanding of the early evolution and current functions of the transcription–translation mechanisms.

Dr. Crews's "Specific Aims"

Version 1

I propose to work in the area of reproductive biology, concentrating on the regulation of reproduction by internal and external stimuli in seasonally breeding vertebrates. Specifically, I will continue my studies of two reptile species that differ markedly in their reproductive physiology.

The green anole lizard is similar to many laboratory and domesticated mammals and birds in that the peak gonadal activity (gamete maturation accompanied by a substantial increase in the circulating level of sex steroid hormones) is *associated* with mating. Species that exhibit such a reproductive tactic frequently have a functional association between sex hormones and sexual behavior. Previous research with the green anole lizard has shed new light on ecological and evolutionary adaptations of the neuroendocrine mechanisms controlling sexual behavior and reproductive physiology. In contrast, the red-sided garter snake, as well as many other vertebrates including some mammals, exhibits a *dissociated* reproductive tactic. In these species, production of gametes and maximal sex hormone secretion are temporally dissociated from mating behavior. In the garter snake, gametes are produced in late summer only after the breeding season is ended; the gametes are then stored until the next mating period. Thus, unlike those species with associated reproductive tactics, mating in the red-sided garter snake occurs when the gonads are completely regressed and circulating levels of sex hormones are low. This implies that the causal mechanisms of mating behavior, at least at the physiological level, must be fundamentally different in species with dissociated reproductive tactics. Recent studies of the red-sided garter snake indicate that this is the case.

The observation that gonadal and behavioral cycles can be dissociated is itself not new, but the implications of this observation have not been fully appreciated. I present here a systematic and comparative series of studies that will focus on specific questions involving the causal mechanisms and functional outcomes of sexual behavior in these two species. From this comparison will emerge a new perspective on the many species, life histories, and sex differences observed in vertebrates. In addition to contributing to our understanding of related areas of reproductive biology, including gamete storage and animal husbandry, this research will yield insight into fundamental reproductive processes.

Version 2

In general I am interested in the biopsychology of reproduction, or more precisely the regulation of reproduction by internal and external stimuli in seasonally breeding invertebrates. The general objectives of my research are to i) investigate how the environment regulates reproduction, ii) determine how reproductively relevant stimuli are perceived and integrated in the central nervous system, iii) demonstrate how the information regulates internal reproductive state, and iv) examine how changes in internal state influence the expression of behavior. To this end, I use a comparative approach that combines and integrates the physiological, morphological, organismal, and ecological levels of analysis. The emphasis on laboratory and field experiments reveals the causal mechanisms and functional outcomes of reproductive behavior on each level without obscuring the relations among the levels. Moreover, the laboratory and field studies are complementary. The field has proven to be a valuable testing ground for hypotheses; similarly, the laboratory is the only possible arena for determining the physiological bases of phenomena observed in the field.

The specific objective is to examine the causal mechanisms and functional outcomes of the two major annual reproductive tactics—associated and dissociated—exhibited by higher vertebrates. In many seasonally breeding vertebrates, gamete production and maximum secretion of sex steroid hormones precedes immediately or coincides with courtship and copulatory (mating) behavior. This annual pattern may be termed the *associated* reproductive tactic, or prenuptial gametogenesis (Figure 1). A markedly different annual pattern is exhibited in many vertebrates, including some mammals, in which the gametes are produced only after the breeding season has ended; the gametes are then stored until the next breeding period. In these species, mating occurs when the gonads are not producing gametes and blood levels of sex steroid hormones are basal. This pattern may be referred to as a *dissociated* reproductive tactic, or postnuptial gametogenesis (figure 1 [Crews's figure no.]).

I will focus on one representative species of each reproductive tactic. The green anole lizard is similar to many laboratory and domesticated mammals and birds in showing the associated tactic. In contrast, the red-sided garter snake shows the dissociated pattern. In many instances a direct comparison of these two species will be made, whereas in other instances gaps in our knowledge must be filled before conceptually valid comparisons can be made. Thus, some of the proposed experiments deal only with one species or tactic. Ultimately, however, my goal is to compare the two tactics at as many levels of organization as are feasible and reasonable. Such a broad approach is crucial if important generalities underlying reproductive processes are ever to emerge. My proposed studies will contribute directly to our understanding of related areas of reproductive biology, including gamete storage and animal husbandry.

Appendix 2

Stages of Revision

1. Both authors circulate among their colleagues drafts that are not submitted for publication.
2. Dr. Bloch submits a manuscript to *Nature.* It is returned by the editor without review.
   Dr. Crews submits a manuscript to *Science.* It is reviewed by two referees, who are split in their decisions, and rejected.
3. Dr. Bloch resubmits a slightly revised version to *Nature,* with a cover letter asking for a review. It is reviewed by three referees and rejected.
   Dr. Crews resubmits a revised version to *Science.* It is reviewed by two more referees, who also split widely, and rejected.
4. Dr. Bloch submits a revised version to *Science.* It is reviewed by two referees who are ambivalent but generally favorable, and rejected.
   Dr. Crews submits a heavily revised manuscript to *Nature.* It is returned by the editor without review.
5. Dr. Bloch submits a revised version to the *Journal of Molecular Evolution.* It is accepted, conditional on changes suggested by two referees and the editor.
   Dr. Crews submits the unrevised *Nature* manuscript to *Proceedings of the National Academy of Sciences.* It is rejected.
6. Dr. Bloch's revised manuscript is accepted at the *Journal of Molecular Evolution* and appears in the December 1983 issue.
   Dr. Crews's unrevised manuscript is accepted at *Hormones and Behavior* on the basis of its previous reviews, and appears in the March 1984 issue.

Scope of Claims

Dr. Bloch

I. "Transfer of control . . . , given the name 'surrogation,' marks the appearance of new kinds of behavior at every level of organization and process, including evolution itself."

II. "A primordial tRNA produces through successive rounds of elongation a molecule with multiple functions of gene, message, and scaffolding, and which serves as the source of the original tRNAs and rRNAs."
III. "The patterns and distributions of homologies make phylogenetic relatedness a more plausible explanation than evolutionary convergence."
IV. "The existence of homologous sequences of tRNAa and 16S rRNA is demonstrated."
V. The sequence of one tRNA is . . .

Dr. Crews

I. Environmental factors influence the evolution and development of three aspects of reproduction: "(i) the functional association between gamete production, sex hormone secretion, and mating behavior, (ii) the functional association between gonadal sex . . . and behavioral sex, (iii) the functional association among the components of sexuality."
II. Environmental factors may cause gamete production, sex hormone secretion, and mating behavior to be dissociated.
III. Gamete production, sex hormone secretion, and mating behavior are dissociated in some species of each class of vertebrates.
IV. Gamete production, sex hormone secretion, and mating behavior are dissociated in the red-sided garter snake.
V. The red-sided garter snake mates at the beginning of warm weather, when sex hormone levels are low.

Comparisons of Review Articles by Dr. Crews

1. *Science*, 1975
2. Manuscript, 1983
3. *Hormones and Behavior*, 1984

Titles

1. "Psychobiology of Reptilian Reproduction"
2. "New Concepts in Behavioral Endocrinology"
3. "Gamete Production, Sex Hormone Secretion, and Mating Behavior Uncoupled"

Opening Sentences

1. "The interaction of behavioral, endocrinological, and environmental factors regulating reproduction has been the subject of intensive investigation in recent years."
2. "Much of the information on the causal mechanisms of vertebrate reproductive behavior has been gathered on highly inbred stocks of rodents and birds living under artificial conditions. . . . Some of the organismal concepts that have emerged are overly narrow and sometimes unrealistic."
3. "A common observation for seasonally breeding vertebrates is that the reproductive processes of gamete production, sex hormone secretion, and mating behavior coincide, and further, that sex steroid hormones activate mating behavior. The postulate of hormone dependence of mating behavior is based primarily on detailed studies of laboratory and domesticated species."

Concluding Sentences

1. "Thus, while the utilization of inbred species contributes greatly to our understanding of the factors regulating reproduction, the integration of these factors can only be appreciated fully in an ecological context where the adaptive significance of such interactions becomes apparent."
2. With this work on the *Cnemidophorus*, "it becomes possible to apply evolutionary theory to gain insight into the evolution of psychoneuroendocrine controlling mechanisms."
3. "The possibility that similarities in mating behavior in different vertebrate species is the result of convergent, rather than divergent, evolution adds another perspective to our understanding of the many species, life history, and sex differences observed in vertebrates."

Length

1. 3200 words
2. 2992 words
3. 940 words

Examples of Revisions

Dr. Bloch

A1. "The Evolution of Control Systems: The Evolution of Evolution"
A2. "An Argument for a Common Evolutionary Origin of tRNAs and rRNAs"
A3. "tRNA–rRNA sequence Homologies: Evidence for a Common Evolutionary Origin?"

B1. ". . . peppered with stretches . . ."
B2. ". . . were found to contain stretches . . ."

C1. Heading "Why the Homologies?"
C2. "Discussion"

D1. The role of coincidence in some matches "will be revealed . . ."
D2. It "should be revealed . . ."

E1. "This is a tantalizing bit of numerology that evokes no ready explanation from current views of RNA functions or relationships."
E2. "This interesting stoichiometry . . ."
E3. "This suggestion of a stoichiometry . . ."

Dr. Crews

A1. "New Concepts in Behavioral Endocrinology"
A2. "Functional Association in Behavioral Endocrinology: Gamete Production, Sex Hormone Secretion, and Mating Behavior"
A3. "Gamete Production, Sex Hormone Secretion, and Mating Behavior Uncoupled"

B1. "This survey makes several points . . ."
B2. "This survey raises several questions . . ."

C1. "My laboratory has been investigating . . ."
C2. "The most thoroughly investigated species is. . . "

D1. 57 references
D2. 195 references
D3. 52 references

Appendix 3: Figures from Crews and Fitzgerald 1980

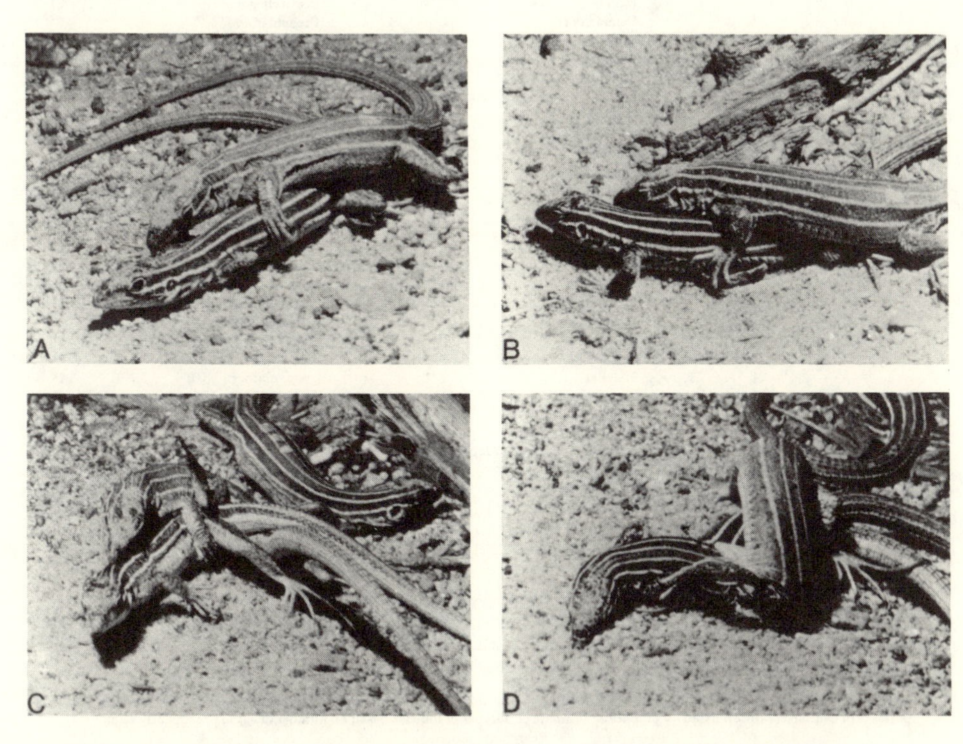

FIG. 1. "Sexual" behavior in captive parthenogenetic *Cnemidophorus uniparens.* After lunging attacks directed at the smaller female, the larger female approaches the now passive small female, first gripping in her jaws the foreleg (*A*). This is accompanied by mounting and riding behavior (*A, B*), during which the active female scratches the side of the mounted female with her fore- and hind-legs and strokes the back of her neck with her jaw. Shortly afterwards, the active female twists her tail beneath the other's tail (*C*), apposing the cloacae and assuming the copulatory posture characteristic of sexual cnemidorphorine lizards (*D*). Females were housed in pairs or groups in aquaria measuring 76.2 × 30.5 cm. Heat was provided by a 75-W, 120-V lamp suspended 10 cm from the sand substrate. A water bowl was provided at the opposite end of the cage. Each cage was illuminated by two Durotest Vita lights 30 cm above the cage bottom. A 14-hr dark: 10-hr light cycle was employed, with a daily temperature gradient of 25°C near the water dish and 47°C directly under the heat lamp. The temperature dropped to 21°C at night. Lizards were fed both mealworms and crickets ad lib. Further details of care and maintenance procedures of parthenogenetic *Cnemidophorus* are provided in ref. 6.

Figure A3.1. The narrative of *Cnemidophorus* behavior in illustrations. From " 'Sexual' Behavior in Parthenogenetic Lizards (*Cnemidophorus*)" by David Crews and Kevin T. Fitzgerald, *Proceedings of the National Academy of Sciences*, Vol. 77, No. 1, January 1980, p. 500. Reproduced by permission of David Crews.

Table 1. Reproductive condition of three species of parthenogenetic lizards (*Cnemidophorus uniparens*, *C. velox*, and *C. tesselatus*) at time of observations

Pair	Animal	Behavior	Snout-vent length, mm	Ovarian condition and size of largest follicles, mm	Number of follicles
			Cnemidophorus uniparens		
A	1	Femalelike	68	Preovulatory (4.8–5.4)	7*
	2	Malelike	59	Previtellogenic (1.2–2.2)	7*
B	3	Femalelike	57	Preovulatory (6.0–6.2)	7*
	4	Malelike	67	Previtellogenic	5†
C	5	Femalelike	60	Preovulatory	3†
	6	Malelike	56	Postovulatory	0†
D	7	Femalelike	67	Preovulatory	4†
	8	Malelike	56	Postovulatory	0†
E	9	Femalelike individual not identified			
	10	Malelike	65	Postovulatory	2†
F	11	Femalelike	66	Preovulatory	3†
	12	Malelike	57	Postovulatory	2†
G	13	Femalelike	63	Preovulatory	2†
	14	Malelike	57	Postovulatory	2†
H	15	Femalelike	66	Preovulatory (≥6.0)	3‡
	16	Malelike	68	Postovulatory (≤3.0)	3‡
I	17	Femalelike	69	Preovulatory (≥6.0)	3‡
	18	Malelike	71	Postovulatory (≤3.0)	2‡
J	19	Femalelike	65	Preovulatory (≥6.0)	3‡
	20	Malelike	71	Postovulatory (≤3.0)	2‡
K	21	Femalelike	66	Preovulatory (≥6.0)	3‡
	22	Malelike	71	Postovulatory (≤3.0)	2‡
L	23	Femalelike	66	Preovulatory (≥6.0)	3‡
	24	Malelike	72	Postovulatory (≤3.0)	2‡
M	25	Femalelike	66	Preovulatory (≥6.0)	3‡
	26	Malelike	71	Postovulatory (≤3.0)	3‡
			Cnemidophorus velox		
A	1	Femalelike	58	Preovulatory (6.5–7.0)	4*
	2	Malelike	67	Previtellogenic (1.2–2.0)	7*
B	3	Femalelike	69	Preovulatory (6.5–7.5)	6*
	4	Malelike	55	Previtellogenic (1.2–1.4)	6*
C	5	Femalelike	66	Preovulatory (6.4–7.0)	5*
	6	Malelike	63	Previtellogenic (0.8–1.0)	5*
			Cnemidophorus tesselatus		
A	1	Femalelike	75	Preovulatory (7.5–8.0)	5*
	2	Malelike	69	Previtellogenic (1.2–2.0)	8*

Female reproductive state was determined by dissection at the time of the observations, egg-laying records, or palpation as noted.
* Determined by immediate dissection.
† Estimate of reproductive condition based on egg-laying record or, in the instance of females 6 and 8, on change in body weight.
‡ Estimate of ovarian condition based on palpation; estimate of number of follicles based on number of eggs subsequently laid.

Figure A3.2. The narrative of *Cnemidophorus* behavior in a table. From " 'Sexual' Behavior in Parthenogenetic Lizards (*Cnemidophorus*)" by David Crews and Kevin T. Fitzgerald, *Proceedings of the National Academy of Sciences*, Vol. 77, No. 1, January 1980, p. 501. Reproduced by permission of David Crews.

Appendix 4

Comparisons of Scientific American *and* New Scientist *Articles with Articles in* Science *or* Evolution

Articles for Professional Audiences	*Articles for Popular Audiences*
1. K. Williams and L. Gilbert, *Science*, 1980	1. L. Gilbert, *Scientific American*, 1982
2. W. Garstka and D. Crews, *Science*, 1981	2. D. Crews and W. Garstka, *Scientific American*, 1982
3. G. Parker, *Evolution*, 1974	3. G. Parker, *New Scientist*, 1979
Titles	
1. "Insects as Selective Agents on Vegetative Morphology: Egg Mimicry Reduces Egg Laying by Butterflies"	1. "Coevolution of a Butterfly and a Vine"
2. "Female Sex Pheromeone in the Skin and Circulation of a Garter Snake"	2. "The Ecological Physiology of a Garter Snake"
3. "The Reproductive Behavior and the Nature of Sexual Selection in *Scatophaga stercoraria* L. (Diptera: Scatophagidae). IX. Spatial Distribution of Fertilization Rates and Evolution of Male Search Strategy Within the Reproductive Area."	3. "Sex Around the Cow-pats"
Abstracts	
1. "Experiments show that *Heliconius* butterflies are less likely to oviposit on host plants that possess eggs or egg-like structures. The egg mimics are an unambiguous example of a plant trait evolved in response to a	1. "*Heliconius* butterflies lay their eggs only on *Passiflora* vines. In defense the vines seem to have evolved fake eggs that make it look to the butterflies as if eggs have already been laid on them."

host restricted group of insect herbivores."

2. "Serums and extracts of tissue from the female garter snake (*Thamnophis sirtalis parietalis*) each act as a pheromone and elicit male courtship behavior when applied to the back of another male. Since pheromonal activity is present in the yolk and liver tissue of untreated females and can be induced with estrogen treatment in serums and livers of males, the pheromone may be associated with circulating yolk lipoprotein, vitellogenin."

2. "In order to survive the harsh environment of western Canada the red-sided garter snake has evolved a precisely-timed cycle of physiology and behavior with several spectacular features."

3. No abstract.

3. "Careful observation of dungflies as they mate on and around cowpats reveals that they use sophisticated strategies in maximising their reproductive success."

Introductions

1. "The idea of coevolution between insects and plants is attractive to biologists attempting to account for patterns of plant chemistry and the use of plants by insects (1). However, it is difficult to demonstrate a causal connection between a plant characteristic and a particular selective agent."

1. "Perhaps the most significant category of ecological interactions in terms of the net transfer of energy in the global food web is the interactions between plants and animals."

2. "In many vertebrates, urine, feces, and vaginal contents, as well as exocrine glandular products, function as sex attractants and serve to facilitate the location and recognition of males (1). We now report an additional source for a vertebrate sex hormone."

2. "The red-sided garter snake is found further north than any other reptile in the Western Hemisphere. It ranges into western Canada, where the winter temperature is often below −40° Celsius and the snow cover often continuous from late September through May.

3. "The present series of papers is aimed towards constructing a comprehensive model of sexual selection and its influence on reproductive strategy in the dungfly, *Schatophaga stercoraria*. The technique used links ecological and behavioral data obtained in the field with laboratory data on sperm competition, for which a model has already been developed (Parker, 1970a)."

3. "Why do peacocks sport outrageously resplendent plumage compared with their more conservative mates? Why do majestic red deer stags engage in ferocious combat with each other for possession of harems, risking severe injury from their spear-point antlers?"

Organization

1. Experiment/control comparisons: "Plants without egg mimics seemed to be more satisfactory for oviposition than plants with egg mimics."

2. Since/then arguments: "Since the female attractiveness pheromone is present in the liver, but not in the fat bodies, and since estrogen treatment can induce the pheromone in the liver and serums of males, we suggest that the pheromone is either the lipoprotein vitellogenin or a lipid-rich part of that large molecule."

3. Predicted/observed comparisons: "For equilibrium between grass and dung surface gain rates, the following algorithm can be formulated from the previous sections:

$$\frac{g_i}{\left\{\dfrac{m_p}{f_i\,(1-P_g)}\right\}+46.5}+\frac{g_o}{\left\{\dfrac{57.2m_p}{f_o(.07m_p+.1)}\right\}+42.25}$$

1. Narrative of the butterfly attacking the vine: "first phase," "second phase," "third phase."

2. Narrative of the reproductive cycle of the snake: female attractiveness pheromone, male unattractiveness pheromone, hormonal relations after mating.

3. Narrative of the mating process: arrival of males, guarding of the females, capture of the females, strategies of searching.

$$+\frac{g_c}{\left\{\frac{154m_p}{f_c(1-P_g)(1-P_e)(.07m_p+.1)}\right\}+53.5}$$

$$=\frac{g_i}{\left\{\frac{m_g}{f_iP_g}\right\}+46.5} \quad (1)$$

where the only variables at a given age of dropping are m_p and m_g. For any total number of searching males (m_t) around a dropping, $m_t = m_p + m_g$. Thus at any value of m_t, the model predicts m_p and m_g. This model is reasonably accurate where search time is long relative to gain extraction time, as with the present data."

Examples of the Editing of Gilbert's Scientific American *Manuscript*

1. Gilbert's manuscript
2. Editor's version
3. Gilbert's revision of editor's version

Addition of Narrative Markers

1. To answer this question involves understanding how *Heliconius* find their host plants, how they decide whether or not to leave an egg, and precisely what factors influence the survival of *Heliconius* larvae once they are on the plant.
2. To answer this question one must understand three aspects of interaction between the butterfly and the vine. The *first* is how the female butterfly finds the host plant. The *second* is how the butterfly makes a choice between depositing its eggs or not depositing them. The *third* consists of the factors that affect the survival [of] the eggs and the caterpillars after they are in place on the vine. Natural selection . . . would favor the mutant vine that was harder for the butterfly to *find*, that was less likely to be selected for egg-laying *once it was found*, and

that was inhospitable to the butterfly's offspring *once they were hatched.* [emphasis added]

Introductory Questions

1. 2 questions
2. 10 questions

Single Sentences Rewritten as Several Sentences

1. Because species placing eggs singly are cannibalistic as larvae, females adding eggs to both occupied and unoccupied shoots at random will leave less progeny than females possessing egg avoiding behaviors.
2. To consider predation, the emerging caterpillars of most *Heliconius* species that deposit single eggs are cannibalistic. One may suppose, then, that a major criterion affecting the decision of a female of these species to deposit eggs or not to deposit eggs would be the presence of another female's eggs at the selected site. A mechanism favoring the avoidance of such sites could easily evolve because mutant butterflies with such a mechanism will have more numerous progeny than butterflies the deposit eggs at occupied and unoccupied sites randomly.

Passive Constructions Rewritten as Active

1. When branches of the host plant having similar oviposition sites were placed in the area, no investigations were made by the *H. hewitsoni* females.
2. I collected lengths of *P. pittieri* vines with newly developed shoots and placed them in the patch of vines that was being regularly visited. The females did not, however, investigate the potential egg-laying sites I had supplied.

1. The observation that inexperienced females are strongly attracted to wire models of tendrils . . . suggests . . .
2. For example [WHAT INVESTIGATOR OF WHAT INSTITUTION?], working with *Heliconius* females in the laboratory, showed that they were strongly attracted to wire models of passion vine tendrils. This behavior suggests that . . .
3. Studies of young, inexperienced *Heliconius* females carried out by Peter Abrams in my laboratory, showed that . . .

Terms (emphasis added)

1. *oviposition*
2. *egg-laying*

1. *Germination,* and therefore small plants, occur[s] in forest gaps where disturbances such as treefalls and landslips have exposed the soil to sunlight.
2. A passion vine seed can *germinate* only on open ground where the soil is exposed to sunlight.

1. . . . *divergence* in the visual appearance of *sympatric* vine species . . .
2. Where different passion vine *species coexist* they *differ* from one another in leaf shape.

Added transitions

2. This suggests, of course, that the pressure of *Heliconius* parasitism has favored the evolution of passion vine leaves to deceive the female butterfly.
3. This suggests that the pressure . . .

2. This being self-evident . . .
3. This being the case . . .
   [Marginal note: "It sure isn't self-evident until you make the observations . . ."]

Appendix 5

From E. O. Wilson, Sociobiology, *p. 550*

How can such variation in social structure persist? The explanation may be lack of competition from other species, resulting in what biologists call ecological release. During the past ten thousand years or longer, man as a whole has been so successful in dominating his environment that almost any kind of culture can succeed for a while, so long as it has a modest degree of internal consistency and does not shut off reproduction altogether. No species of ant or termite enjoys this freedom. The slightest inefficiency in constructing nests, in establishing odor trails, or in conducting nuptial flights could result in the quick extinction of the species by predation and competition from other social insects. To a scarcely lesser extent the same is true for social carnivores and primates. In short, animal species tend to be tightly packed in the ecosystem with little room for experimentation or play. Man has temporarily escaped the constraint of interspecific competition. Although cultures replace one another, the process is much less effective than interspecific competition in reducing variance.

It is part of the conventional wisdom that virtually all cultural variation is phenotypic rather than genetic in origin. This view has gained support from the ease with which certain aspects of culture can be altered in the space of a single generation, too quickly to be evolutionary in nature. The drastic alteration in Irish society in the first two years of the potato blight (1846–1848) is a case in point. Another is the shift in the Japanese authority structure during the American occupation following World War II. Such examples can be multiplied endlessly—they are the substance of history. It is also true that human populations are not very different from one another genetically. When Lewontin (1972b) analyzed existing data on nine blood-type systems, he found that 85 percent of the variance was composed of diversity within populations and only 15 percent due to diversity between populations. There is no a priori reason for supposing that this sample of genes possesses a distribution much different from those of other, less accessible systems affecting behavior.

The extreme orthodox view of environmentalism goes further, holding that in effect there is no genetic variance in the transmission of culture. In other words, the capacity for culture is transmitted by a single human genotype. Dobzhansky (1963) stated this hypothesis as follows: "Culture is not inherited through genes, it is acquired by learning from other human beings . . . In a sense, human genes have surrendered their primacy in human evolution to an entirely new, nonbiological or superorganic agent, culture. However, it should not be forgotten that this agent is entirely dependent on the human genotype." Although the genes have given away most of their sovereignty, they maintain a certain amount of influence in at least the behavioral qualities that underlie variations between cultures. Moderately high heritability has been documented in introversion-extroversion measures, personal tempo, psychomotor and sports activities, neuroticism, dominance, depression, and the tendency toward certain forms of mental illness such as schizophrenia (Parsons, 1967; Lerner, 1968). Even a small portion of this variance invested in population differences might predispose societies toward cultural differences. At the very least, we should try to measure this amount. It is not valid to point to the absence of a behavioral trait in one or a few societies as conclusive evidence that the trait is environmentally induced and has no genetic disposition in man. The very opposite could be true.

In short, there is a need for a discipline of anthropological genetics. In the interval before we acquire it, it should be possible to characterize the human bioprogram by two indirect methods. First, models can be constructed from the most elementary rules of human behavior. Insofar as they can be tested, the rules will characterize the biogram in much the same way that ethograms drawn by zoologists identify the "typical" behavioral repertories of animal species. The rules can be legitimately compared with the ethograms of other primate species. Variation in the rules among human cultures, however slight, might provide clues to underlying genetic differences, particularly when it is correlated with variation in behavioral traits known to be heritable. Social scientists have in fact begun to take this first approach, although in a different context from the one suggested here. Abraham Maslow (1954, 1972) postulated that human beings respond to a hierarchy of needs, such that the lower levels must be satisfied before much attention is devoted to the higher ones. The most basic needs are hunger and sleep. When these are met, safety becomes a primary consideration, then the need to belong to a group and receive love, next self-esteem, and finally self-actualization and creativity. The ideal society in Maslow's dream is one which "fosters the fullest development of human potentials, of the fullest degree of humanness." When the biogram is freely expressed, its center of gravity should come to rest in the higher levels. [*Sociobiology: The New Synthesis*, E. O. Wilson, Cambridge: Harvard University Press, 1975. Copyright 1975 by the President and Fellows of Harvard College. Reprinted by permission.]

References

1. Texts Cited

Abir-Am, Pnina. "Themes, Genres, and Orders of Legitimation in the Consolidation of New Scientific Disciplines: Deconstructing the Historiography of Molecular Biology." *History of Science* 23 (1985): 73–117.

Agassiz, Elizabeth Cary. *Louis Agassiz: His Life and Correspondence.* 2 vols. Boston: Houghton Mifflin, 1886.

Albury, W. R. "Politics and Rhetoric in the Sociobiology Debate." *Social Studies of Science* 10 (1980): 519–36.

Ashmore, Malcolm, Michael Mulkay, and Trevor Pinch. *Health and Efficiency.* Milton Keynes: Open University Press, 1989.

Barker, Martin. *The New Racism.* London: Junction, 1981.

Barker, Martin. "Sociobiology and Ideology: The Unfinished Trajectory." Unpublished MS, 1986, Bristol Polytechnic, Department of Philosophy.

Barnes, Barry. *About Science.* Oxford: Basil Blackwell, 1985.

Barnes, Barry. *Scientific Knowledge and Sociological Theory.* London: Routledge and Kegan Paul, 1974.

Barnes, Barry, and David Edge, eds. *Science in Context: Readings in the Sociology of Science.* Milton Keynes: Open University Press, 1982, and Cambridge: MIT Press, 1985.

Bastide, François. "The Semiotic Analysis of Discourse." Unpublished MS, Paris: École des Mines, 1986.

Bastide, François. "Une Nuit avec Saturne." Unpublished MS, Paris: École des Mines, 1986.

Bazerman, Charles. *The Inquiring Writer.* Boston: Houghton Mifflin, 1981; 1985, 1989.

Bazerman, Charles. "Modern Evolution of the Experimental Report in Physics: Spectroscopic Articles in *Physical Review*, 1893–1980." *Social Studies of Science* 14 (1984): 163–96.

Bazerman, Charles. "Physicists Reading Physics: Schema-Laden Purposes and Purpose-Laden Schema." *Written Communication* 2 (1985): 3–23.

Bazerman, Charles. "Scientific Writing as a Social Act: A Review of the Literature of the Sociology of Science." In *New Essays in Technical and Scientific Communication: Research, Theory, Practice*, eds. Paul V. Anderson, R. John Brockmann, and Carolyn R. Miller. Farmingdale, N.Y.: Baywood, 1982. Pp. 156–84.

Bazerman, Charles. *Shaping Written Knowledge: Essays in the Growth, Form, Function, and Implications of the Scientific Article.* Madison: University of Wisconsin Press, 1988.

Bazerman, Charles. "Studies of Scientific Writing—E Pluribus Unum?" *4S Review* 3, no. 2 (1985): 13–20.

Bazerman, Charles. "What Written Knowledge Does: Three Examples of Academic Discourse." *Philosophy of the Social Sciences* 11 (1981): 361–87.

Bazerman, Charles. "The Writing of Scientific Non-Fiction: Contexts, Choices, Constraints." *Pre/Text* 5 (1984): 39–74.

Benjamin, Andrew, Geoffrey Cantor, and J. R. R. Christie, eds., *The Figural and the Literal: Problems of Language in the History of Science and Philosophy.* Manchester: Manchester University Press, 1987.

Beer, Gillian. *Darwin's Plots: Evolutionary Narrative in Darwin, George Eliot, and Nineteenth-Century Fiction.* London: Routledge and Kegan Paul, 1983.

Billig, Michael. *Arguing and Thinking: A Rhetorical Approach to Social Psychology.* Cambridge: Cambridge University Press, 1987.

Birke, Lynda, and Jonathan Silverton, eds. *More than the Parts: Biology and Politics.* London: Pluto Press, 1984.

Bizzell, Patricia. "Cognition, Convention, and Certainty: What We Need to Know About Writing." *Pre/Text* 3 (1982): 215–43.

Bloor, David. *Knowledge and Social Imagery.* London: Routledge and Kegan Paul, 1976.

Brannigan, Augustine. *The Social Basis of Scientific Discoveries.* Cambridge: Cambridge University Press, 1981.

Callon, Michel. "Struggles and Negotiations to Define What Is Problematic and What Is Not: The Socio-logic of Translation." In *The Social Process of Scientific Investigation.* Sociology of the Sciences, Vol. IV, eds. K. Knorr, R. Krohn, and R. Whitley. Dordrecht: Reidel, 1980. Pp. 197–219.

Callon, Michael, and Bruno Latour. "Unscrewing the Big Leviathan: How Actors Macrostructure Reality and Sociologists Help Them to Do So." In *Advances in Social Structure and Methodology: Toward an Integration of Micro and Macro-Sociologies,* eds. K. Knorr-Cetina and A. Cicourel. London: Routledge and Kegan Paul, 1981. Pp. 277–303.

Callon, Michael, John Law, and Arie Rip, eds. *Mapping the Dynamics of Science and Technology: Sociology of Science in the Real World.* London: Macmillan, 1986.

Cannon, Walter F. "Darwin's Vision in *On the Origin of Species.*" In *The Art of Victorian Prose,* eds. George Levine and William Madden. New York: Oxford University Press, 1968. Pp. 154–76.

Chubin, Daryl, and S. Moitra. "Content Analysis of References." *Social Studies of Science* 5 (1975): 423–41.

Clark, Ronald. *J.B.S.: The Life and Work of J.B.S. Haldane.* London: Hodder and Stoughton, 1968.

Clifford, James E., and George E. Marcus, eds. *Writing Culture: The Poetics and Politics of Ethnography.* Berkeley: University of California Press, 1986.

Cloître, Michel, and Terry Shinn. "Enclavement et Diffusion du Savoir." *Social Science Information* 25, no. 1 (1986): 161–87.

Cloître, Michel, and Terry Shinn. "Expository Practice: The Social, Cognitive, and Epistemological Linkage." *Expository Science: Forms and Functions of Popularisation.* Sociology of the Sciences Yearbook, Vol. 9. Dordrecht: D. Reidel, 1985. Pp. 31–60.

Cole, Jonathan, and Stephen Cole. "Which Researcher Will Get the Grant?" *Nature,* 279 (1979): 575–76.

Cole, Stephen, Jonathan R. Cole, and Gary A. Simon. "Chance and Consensus in Peer Review." *Science* 214 (1981): 881–86.

Cole, Stephen, R. Rubin, and Jonathan R. Cole. "Peer Review and the Support of Science." *Scientific American* 237, no. 4 (1977): 34–41.

Collins, Harry. *Changing Order: Replication and Induction in Scientific Practice.* Beverly Hills and London: Sage, 1985.

Collins, Harry. "An Empirical Relativist Programme in the Sociology of Scientific Knowledge." In *Science Observed: Perspectives on the Social Study of Science,* eds. Karin D. Knorr-Cetina and Michael Mulkay. Beverly Hills and London: Sage, 1983. Pp. 85–113.

Collins, Harry. "The Seven Sexes: A Study in the Sociology of a Phenomenon, or the Replication of Experiments in Physics." *Sociology* 9 (1975): 205–24. Rpt. in *Science in Context,* eds. Barry Barnes and David Edge. Milton Keynes: Open University Press, 1982. Pp. 94–116.

Collins, Harry. "Son of the Seven Sexes: The Social Destruction of a Physical Phenomenon." *Social Studies of Science* 11 (1981): 33–62.

Collins, Harry, and Trevor Pinch. *Frames of Meaning: The Social Construction of Extraordinary Science.* London: Routledge and Kegan Paul, 1982.

Conan Doyle, Arthur. *Sherlock Holmes: The Complete Long Stories.* London: John Murray, 1929.

Cooper, Charles, and Lee Odell. "Considerations of Sound in the Composing Processes of Published Writers." *Research in the Teaching of English* 10 (1976): 103–15.

Coulthard, Malcolm. *An Introduction to Discourse Analysis.* London: Longman, 1977.

Cozzens, Susan. "Comparing the Sciences: Citation Context Analysis of Papers from Neuropharmacology and the Sociology of Science." *Social Studies of Science* 15 (1985): 127–53.

Crocker, D. R. "Anthropomorphism: Bad Practice, Honest Prejudice?" *New Scientist* 16 July (1981): 159–62.

Crocker, Joe. "Sociobiology: The Capitalist Synthesis." *Radical Science Journal* 13 (1983): 55–72.

Culler, Dwight. "The Darwinian Revolution and Literary Form." In *The Art of Victorian Prose*, eds. George Levine and William Madden. New York: Oxford University Press, 1968. Pp. 224–46.

Darwin, Charles. *Journal of Researches into the Natural History and Geology of the Countries Visited During the Voyage Round the World of H.M.S. 'Beagle' Under Command of Captain Fitz Roy, R. N.* 1845; reprinted London: John Murray, 1901.

Darwin, Charles. *The Origin of Species*. 1859; reprinted Harmondsworth: Penguin, 1968.

Dawkins, Richard. *The Selfish Gene*. Oxford University Press, 1976.

Dialectics of Biology Group. *Against Biological Determinism*. London: Allison and Busby, 1982.

Dudley-Evans, Tony, ed. "Genre Analysis and ESP." *ELR Journal* 1 (1987).

Eagleton, Terry. *Literary Theory: An Introduction*. Oxford: Blackwell, 1984.

Eisenstein, Elizabeth K. *The Printing Press as an Agent of Change*. Cambridge: Cambridge University Press, 1979.

Ezrahi, Yaron. "The Authority of Science in Politics." In *Science and Values*, eds. A. Thackray and E. Mendelsohn. New York: Humanities Press, 1974. Pp. 215–51.

Fleck, Ludwik. *The Genesis and Development of a Scientific Fact*. Chicago: University of Chicago Press, 1981.

Freud, Sigmund. "From the History of an Infantile Neurosis (The 'Wolf Man')." *The Pelican Freud Library, Volume 9: Case Histories II*. Harmondsworth: Penguin, 1979. Pp. 227–366.

Garfinkel, Harold. *Studies in Ethnomethodology*. Englewood Cliffs: Prentice-Hall, 1967; reprinted, Cambridge: Polity Press, 1984.

Garfinkel, Harold, Michael Lynch, and Eric Livingston. "The Work of a Discovering Science Construed with Materials the Optically Discovered Pulsar." *Philosophy of the Social Sciences* 11 (1981): 131–58.

Gilbert, Nigel, and Michael Mulkay, "Experiments Are the Key: Participants' Histories and Historians' Histories of Science." *Isis* 75 (1984): 105–25.

Gilbert, Nigel, and Michael Mulkay. *Opening Pandora's Box: A Sociological Analysis of Scientists' Discourse*. Cambridge: Cambridge University Press, 1984.

Gilbert, Nigel. "The Transformation of Research Findings into Scientific Knowledge." *Social Studies of Science* 6 (1976): 281–306.

Golinski, Jan. "Robert Boyle: Skepticism and Authority in Seventeenth-

Century Chemical Discourse." In *The Figural and the Literal: Problems of Language in the History of Science and Philosophy*, eds. Andrew Benjamin, Geoffrey Cantor, and J. R. R. Christie. Manchester: Manchester University Press, 1987. Pp. 58–82.

Gordon, M. "The Role of Referees in Scientific Communication." In *The Psychology of Written Communication: Selected Readings*, ed. J. Hartley. London: Kogan Page, 1980. Pp. 263–75.

Gossler, M. "Numerology." *Nature* 306 (1983): 530.

Haldane, J. B. S. *The Inequality of Man and Other Essays*. Harmondsworth: Penguin, 1937.

Haldane, J. B. S. *On Being the Right Size*. Oxford: Oxford UP, 1985.

Halfpenny, Peter. "Talking of Writing, Writing of Writing: Some Reflections on Gilbert and Mulkay's Discourse Analysis." *Social Studies of Science* 18 (1988): 169–82.

Heritage, John. *Garfinkel and Ethnomethodology*. Cambridge: Polity Press, 1984.

Hesse, Mary. *Revolutions and Reconstructions in the Philosophy of Science*. Brighton: Harvester, 1980.

Holmes, Frederick L. "Lavosier and Krebs: The Individual Scientist in the Near and Deeper Past." *Isis* 75 (1984): 131–42.

Holmes, Frederick L. "Writing and Discovery." *Isis* 78 (1987): 220–35.

Hull, David L. "Openness and Secrecy in Science: Their Origins and Limitations." *Science, Technology, and Human Values* 10, no. 2 (1985): 4–13.

Hull, David L. "Thirty-one Years of *Systematic Zoology*." *Systematic Zoology* 32, no. 4 (1983): 315–42.

Jacobi, Daniel. "Références Iconiques et Modèles Analogiques dans des Discours de Vulgarisation Scientifique." *Social Science Information* 24, no. 4 (1985): 847–67.

Jameson, Fredric. *The Political Unconscious: Narrative as a Socially Symbolic Act*. Ithaca: Cornell University Press, 1981.

Kenward, M. "Peer Review and the Axe Murderers." *New Scientist* 31 May (1984): 13.

Knorr-Cetina, Karin. *The Manufacture of Knowledge: An Essay on the Constructivist and Social Nature of Science*. Oxford: Pergamon, 1981.

Knorr-Cetina, Karin, and Michael Mulkay, eds. *Science Observed: Perspectives on the Social Study of Science*. Beverly Hills and London: Sage, 1983.

Landau, Misia. "Human Evolution as Narrative." *American Scientist* 72 (1984): 262–67.

Latour, Bruno. "Give Me a Laboratory and I Will Raise the World." In *Science Observed*, eds. Karin D. Knorr-Cetina and Michael Mulkay. Beverly Hills and London: Sage, 1983. Pp. 141–70.

Latour, Bruno. *Les Microbes: Guerre et Paix*. Paris: Éditions A. M. Métailie,

1984. Translated as *The Pasteurization of France*. Cambridge: Harvard University Press, 1988.

Latour, Bruno. "A Relativistic Account of Einstein's Relativity." Unpublished MS, École des Mines, Paris, 1986.

Latour, Bruno. *Science in Action: How to Follow Scientists and Engineers Through Society*. Milton Keynes: Open UP, 1987.

Latour, Bruno, and Françoise Bastide. "Writing Science—Fact and Fiction: The Analysis of the Process of Reality Construction Through the Application of Socio-Semiotic Methods to Scientific Texts." In *Mapping the Dynamics of Science and Technology*, eds. Michael Callon, John Law, and Arie Rip. London: Macmillan, 1986. Pp. 51–67.

Latour, Bruno, and Jocelyn de Noblet, eds. *Les "Vues" de L'Esprit*, special issue of *Culture Technique* 14 (Juin 1985).

Latour, Bruno, and S. C. Strum. "Human Social Origins: Oh Please, Tell Us Another Story." *Journal of Biological and Social Structure* 9 (1986): 169–87.

Latour, Bruno, and Steve Woolgar. *Laboratory Life: The Social Construction of Scientific Facts*. Beverly Hills and London: Sage, 1979; 2nd ed., *Laboratory Life: The Construction of Scientific Facts*. Princeton: Princeton University Press, 1986.

Law, John, and Rob Williams. "Putting Facts Together: A Study in Scientific Persuasion." *Social Studies of Science* 12 (1982): 535–58.

Lentricchia, Frank. *After the New Criticism*. Chicago: University of Chicago Press, 1981.

Levidow, Les. "IQ as Ideological Reality." In *Radical Science*, ed. Les Levidow. London: Free Association Books, 1986. Pp. 198–213.

Lynch, Michael. *Art and Artifact in Laboratory Science: A Study in Shop Work and Shop Talk in a Research Laboratory*. London: Routledge and Kegan Paul, 1985.

Lynch, Michael. "Discipline and the Material Form of Images: An Analysis of Scientific Visibility." *Social Studies of Science* 15 (1985): 37–66.

Lynch, Michael. "Technical Work and Critical Inquiry: Investigations in a Scientific Laboratory." *Social Studies of Science* 12 (1982): 499–533.

Lynch, Michael, Eric Livingston, and Harold Garfinkel. "Temporal Order in Laboratory Work." In *Science Observed*, eds. Karin D. Knorr-Cetina and Michael Mulkay. Beverly Hills and London: Sage, 1983. Pp. 205–38.

Lynch, Michael, and Steve Woolgar, eds. *Human Studies: Special Issue on Representation in Science* 11 (July 1988).

Lyotard, Jean-François. *The Post-Modern Condition: A Report on Knowledge*. Manchester: Manchester University Press, 1984.

Macdonell, Diane. *Theories of Discourse: An Introduction.* Oxford: Basil Blackwell, 1986.

Mandel, H. "Funding More NIH Research Grants." *Science* 221 (1983): 338–40.

Marcus, George E., and Michael M. J. Fischer. *Anthropology as Cultural Critique: An Experimental Moment in the Human Sciences.* Chicago: University of Chicago Press, 1986.

Mayr, Ernst. *The Growth of Biological Thought: Diversity, Evolution, and Inheritance.* Cambridge, Mass.: Harvard University Press, 1982.

McKinlay, Andrew, and Potter, Jonathan. "Model Discourse: Interpretive Repertoires in Scientists' Conference Talk." *Social Studies of Science* 17 (1987): 443–63.

Medawar, Peter. "Is the Scientific Paper Fraudulent?" *Saturday Review* 49 (1 August 1964): 42–43.

Merton, Robert. *The Sociology of Science.* Ed. Norman Storer. Chicago: University of Chicago Press, 1973.

Miller, Carolyn. "Public Knowledge in Science and Society." *Pre/Text* 3 (1982): 31–49.

Morrison, Kenneth. "Some Properties of Telling-Order Designs in Didactic Inquiry." *Philosophy of the Social Sciences* 11 (1981): 245–62.

Mulkay, Michael. *Science and the Sociology of Knowledge.* London: George Allen and Unwin, 1979.

Mulkay, Michael. *The Word and the World.* London, Allen and Unwin, 1985.

Mulkay, Michael, Malcolm Ashmore, and Trevor Pinch. "Colonising the Mind." Paper presented at ESRC Seminar, York University, September 1986.

Mulkay, Michael, Malcolm Ashmore, and Trevor Pinch. "Measuring the Quality of Life: A Sociological Invention Concerning the Application of Economics to Health Care." *Sociology* 21 (1987): 541–64.

Mulkay, Michael, Jonathan Potter, and Steven Yearley. "Why an Analysis of Scientific Discourse is Needed." In *Science Observed,* eds. Karin D. Knorr-Cetina and Michael Mulkay. Beverly Hills and London: Sage, 1983. Pp. 171–204.

Myers, Greg. "Every Picture Tells a Story: Illustrations in E. O. Wilson's *Sociobiology.*" *Human Studies* 11 (1988): 235–69.

Myers, Greg. "Making a Discovery: Narratives of Split Genes." In *Narrative and Cognition,* ed. Christopher Nash. London: Routledge and Kegan Paul, in press.

Myers, Greg. "Nineteenth-Century Popularizations of Thermodynamics and the Rhetoric of Social Prophecy." *Victorian Studies* 29 (1985): 35–66.

Myers, Greg. "The Pragmatics of Politeness in Scientific Articles." *Applied Linguistics* 10 (1989): 1–35.

Myers, Greg. "Reporting Genetic Fingerprints." Paper presented at the Oxford-Ciba International Symposium on Reporting Science. Oxford, July 1988.

Myers, Greg. "Science for Women and Children: The Dialogue of Popular Science in the Nineteenth Century." In *Nature Transfigured: Literature and Science 1700–1900,* eds. Sally Shuttleworth and J. R. R. Christie. Manchester: Manchester University Press, 1989.

Myers, Greg. "Story and Style in Two Review Articles." In *Textual Dynamics of the Professions,* eds. Charles Bazerman and James Paradis. Madison: University of Wisconsin Press, in press.

National Institutes of Health. *Proceedings of the 1983 Meetings of NIH Scientific Review Groups.* Washington, D.C.: Department of Human Services, 1983.

National Institutes of Health. *Application for Public Health Service Grant.* Washington, D.C.: Department of Health and Human Services, 1984.

Newmeyer, Frederick. *Linguistic Theory in America: The First Quarter-Century of Transformational Generative Grammar.* New York: Academic Press, 1980.

Norell, M. "Homology Defined." *Nature* 306 (1983): 530.

Paludi, Michele A., and Lisa A. Strayer. "What's in an Author's Name? Differential Evaluations of Performance as a Function of Author's Name." *Sex Roles* 12 (1985): 353–61.

Peters, Douglas P., and Stephen J. Ceci. "Peer Review Practices of Psychological Journals: The Fate of Published Articles, Submitted Again." *Behavioral and Brain Sciences* 5 (1982): 187–255.

Pickering, Andrew. *Constructing Quarks: A Sociological History of Particle Physics.* Edinburgh: Edinburgh University Press, 1984.

Pinch, Trevor. "Towards an Analysis of Scientific Observation: The Externality and Evidential Significance of Observational Reports in Physics." *Social Studies of Science* 15 (1985): 3–36.

Potter, Jonathan. "Cutting Cakes: A Study of Psychologists' Social Categorizations." *Philosophical Psychology* 1 (1988): 17–32.

Potter, Jonathan. "Testability, Flexibility: Kuhnian Values in Scientists' Discourse Concerning Theory Choice." *Philosophy of the Social Sciences* 14 (1984): 303–30.

Potter, Jonathan, and Margaret Wetherell. "Accomplishing Attitudes: Fact and Evaluation in Racist Discourse." *Text* 8 (1988): 51–68.

Potter, Jonathan, and Margaret Wetherell. *Discourse and Social Psychology: Beyond Attitudes and Behaviour.* Beverly Hills and London: Sage, 1987.

Ree, Jonathan. *Philosophical Tales.* London: Methuen, 1987.

Rose, Hilary, and Steven Rose. "Radical Science and Its Enemies." In *The Socialist Register 1979*, eds. Ralph Miliband and John Saville. London: Merlin, 1979. Pp. 317–34.

Royal Society. *The Public Understanding of Science.* London: The Royal Society, 1985.

Royal Society. "Science is for Everybody." [leaflet] London: The Royal Society, 1985.

Rudwick, Martin. *The Great Devonian Controversy: The Shaping of Scientific Knowledge among Gentlemanly Specialists.* Chicago: University of Chicago Press, 1985.

Sacks, Harvey, Schegloff, Emmanual, and Jefferson, Gail. "A Simplest Systematics for the Organization of Turn-Taking in Conversation." *Language* 50 (1974): 696–735.

Schrödinger, Erwin. *What is Life?* Cambridge: Cambridge University Press, 1956.

Seidel, Gill. "Political Discourse Analysis." *Handbook of Discourse Analysis,* Vol. 4, ed. Teun van Dijk. London: Academic Press, 1985. Pp. 43–60.

Shapin, Steven. "History of Science and its Sociological Reconstructions." *History of Science* 20 (1982): 157–211.

Shapin, Steven. "Phrenological Knowledge and the Social Structure of Early Nineteenth-Century Edinburgh." *Annals of Science* 32 (1975): 219–43.

Shapin, Steven. "The Politics of Observation: Cerebral Anatomy and Social Interests in the Edinburgh Phrenology Disputes." In *On the Margins of Science: The Social Construction of Rejected Knowledge. Sociological Review* Monograph 27, ed. Roy Wallis. Keele: University of Keele, 1979. Pp. 139–78.

Shapin, Steven. "Pump and Circumstance: Robert Boyle's Literary Technology." *Social Studies of Science* 14 (1984): 481–520.

Shapin, Steven. "Talking History: Reflections on Discourse Analysis." *Isis* 75 (1984): 125–28.

Shapin, Steven, and Simon Schaffer. *Leviathan and the Airpump: Hobbes, Boyle, and the Experimental Life.* Princeton: Princeton University Press, 1984.

Shinn, Terry, and Richard Whitley, eds. *Expository Science: Forms and Functions of Popularization.* Dordrecht: Reidel, 1985.

Shuttleworth, Sally. *George Eliot and Nineteenth-Century Science: The Make-Believe of a Beginning.* Cambridge: Cambridge University Press, 1984.

Sinclair, John, and Malcolm Coulthard. *Towards an Analysis of Discourse: The English Used by Teachers and Pupils.* London: Oxford University Press, 1974.

Sperber, Dan, and Deirdre Wilson. *Relevance: Communication and Cognition.* Oxford: Basil Blackwell, 1986.

Stumpf, Walter E. "Peer Review." *Science* 207 (1980): 823–24.

Swales, John. *Aspects of Article Introductions*. Birmingham: Aston University, 1981.

Swales, John. "Citation Analysis and Discourse Analysis." *Applied Linguistics* 7 (1986): 39–56.

Van den Beemt, F., and C. Le Pair. "Appraisal of Peer Review." Paper presented at the meeting of the Society for Social Studies of Science, Blacksburg, Va. November 1983.

Walker, Teri. "Whose Discourse?" In *Knowledge and Reflexivity: New Frontiers in the Sociology of Knowledge*, ed. Steve Woolgar. London and Beverly Hills: Sage, 1988. Pp. 55–79.

Watson, James. *The Double Helix*. Reprint, ed. Gunther Stent. New York: Norton, 1981.

Werskey, Gary. *The Visible College*. London: Allen Lane, 1978.

White, Hayden. *Metahistory: The Historical Imagination in Nineteenth-Century Europe*. Baltimore: Johns Hopkins University Press, 1973.

White, Hayden. *Tropics of Discourse: Essays in Cultural Criticism*. Baltimore: Johns Hopkins University Press, 1978.

Woolgar, Steve. "Discourse and Praxis." *Social Studies of Science* 16 (1986): 309–18.

Woolgar, Steve. "Discovery: Logic and Sequence in a Scientifc Text." In *The Social Process of Scientific Investigation*. Sociology of the Sciences, Vol. IV, eds. K. Knorr, R. Krohn, and R. Whitley. Dordrecht: D. Reidel, 1980. Pp. 239–68.

Woolgar, Steve. "Interests and Explanations in the Social Study of Science." *Social Studies of Science* 11 (1981): 365–94.

Woolgar, Steve. "Irony in the Social Study of Science." In *Science Observed*, eds. Karin D. Knorr-Cetina and Michael Mulkay. Beverly Hills and London: Sage, 1983. Pp. 239–66.

Woolgar, Steve. *Science: The Very Idea*. London: Tavistock, 1988.

Woolgar, Steve. "Writing an Intellectual History of a Scientific Development: The Use of Discovery Accounts." *Social Studies of Science* 6 (1976) 395–422.

Woolgar, Steve, ed. *Knowledge and Reflexivity: New Frontiers in the Sociology of Knowledge*. Beverly Hills and London: Sage, 1988.

Wynne, Brian. "Between Orthodoxy and Oblivion: The Normalization of Deviance in Science." *On the Margins of Science: The Social Construction of Rejected Knowledge. Sociological Review*, Monograph 27, ed. Roy Wallis. Keele: University of Keele, 1979.

Wynne, Brian. "Establishing the Rules of Laws: Constructing Expert Authority." In *Expert Evidence: Interpreting Science in the Law*, eds. Roger Smith and Brian Wynne. London: Routledge and Kegan Paul, 1988.

Wynne, Brian. *Rationality or Ritual? Nuclear Decision-Making and the Windscale Inquiry*. Chalfont St. Giles, Bucks.: British Society for the History of Science, 1982.

Wynne, Brian. "Unruly Technology: Practical Rules, Impractical Discourses and Public Understanding." *Social Studies of Science* 18 (1988): 147–67.

Wynne, Brian, Peter Williams, and Jean Williams. "Cumbrian Hill-Farmers' Views of Scientific Advice." *Evidence to the House of Commons Select Committee on Agriculture: The Chernobyl Disaster and Effects of Radioactive Fallout on the U.K.* Lancaster: Center for Science Studies and Science Policy. University of Lancaster, 1988.

Yearley, Steven. "Representing Geology: Textual Structures in the Pedagogical Presentation of Science." *Expository Science: Forms and Functions of Popularization*. Sociology of the Sciences Yearbook, Vol. IX, eds. Terry Shinn and Richard Whitley. Dordrecht: D. Reidel, 1985. Pp. 79–101.

Yearley, Steven. "Textual Persuasion: The Role of Social Accounting in the Construction of Scientific Arguments." *Philosophy of the Social Sciences* 11 (1981): 409–35.

Young, Robert. *Darwin's Metaphor*. Cambridge: Cambridge University Press, 1986.

Ziman, John. *An Introduction to Science Studies: The Philosophical and Social Aspects of Science and Technology*. Cambridge: Cambridge University Press, 1984.

Zuckerman, Harriet, and Robert Merton. "Institutionalized Patterns of Evaluation in Science." In Robert Merton, *The Sociology of Science*, ed. Norman Storer. Chicago: University of Chciago Press, 1973. Pp. 460–98.

2. *Texts Discussed: Chapter 3*

The Two Articles Discussed in Detail

Bloch, David, B. McArthur, R. Widdowson, D. Spector, R. Guimares, and J. Smith. "tRNA-rRNA Sequence Homologies: Evidence for a Common Evolutionary Origin?" *Journal of Molecular Evolution* 19 (1983): 420–28.

Crews, David. "Gamete Production, Sex Hormone Secretion, and Mating Behavior Uncoupled." *Hormones and Behavior* 18 (1984): 22–28.

Other Works Mentioned for Comparison

Bloch, David. "tRNA-rRNA Sequence Homologies: A Model for the Origin of a Common Ancestral Molecule, and Prospects for Its Reconstruction." *Origins of Life* 14 (1984): 571–78.

Bloch, David, C.-T. Fu, and P. Dean. "DNA and Histone Synthesis Rate

Change During the S. Period in Erlich Ascites Tumor Cells." *Chromosoma* 82 (1981): 611–26.

Bloch, David, Barbara McArthur, and Sam Mirrop. "tRNA-rRNA Sequence Homologies: Evidence for an Ancient Modular Format Shared by tRNAs and rRNAs." *BioSystems* 17 (1985): 209–25.

Crews, David. "Diversity of Hormone-Behavior Controlling Mechanisms," and "Alternative Reproductive Tactics in Reptiles." *BioScience* 33 (1983): 545, 562–67.

Crews, David. "Functional Associations in Behavioral Endocrinology." In *Masculinity/Femininity: Concepts and Definitions,* ed. J. M. Reinisch. Bloomington: Indiana University Press, 1985.

Crews, David, ed. *Psychobiology of Reproductive Behavior: An Evolutionary Perspective.* Englewood Cliffs, N.J.: Prentice-Hall, 1987.

Crews, David. "Psychobiology of Reptilian Reproduction." *Science* 189 (1975): 1059–1065.

Crews, David, and Michael C. Moore. "Evolution of Mechanisms Controlling Mating Behavior." *Science* 231 (1986): 121–25.

Lewin, Roger. "Basic Modular Formats in tRNA's and rRNA's." *Science* 229 (1985): 1254.

Nazarea, Apolinario D., David P. Bloch, and Anne C. Semrau. "Detection of a Fundamental Modular Format Common to Transfer and Ribosomal RNAs: Second-Order Spectral Analysis." *Proceedings of the National Academy of Sciences USA* 82 (1985): 5337–42.

3. Texts Discussed: Chapter 4

Cole, Charles J. "Parthenogenetic Lizards [Technical Comment]." *Science* 201 (1978): 1154–55.

Cole, Charles J. "Unisexual Lizards." *Scientific American* 250 (January 1984): 84–90.

Cole, Charles J., and Carol R. Townsend. "Sexual Behavior in Unisexual Lizards." *Animal Behaviour* 31 (1983): 724–28.

Crews, David, "Courtship in Unisexual Lizards: A Model for Brain Evolution." *Scientific American* 225 (December 1987): 116–21.

Crews, David, and Kevin T. Fitzgerald. " 'Sexual' Behavior in Parthenogenetic Lizards (*Cnemidophorus*)." *Proceedings of the National Academy of Sciences* 77 (1980): 499–502.

Crews, David, Jill E. Gustafson, and Richard E. Tokarz. "Psychobiology of Parthenogenesis." In *Lizard Ecology,* eds. R. B. Huey, E. R. Pianka, and T. W. Schoener. Cambridge: Harvard University Press, 1983. Pp. 205–31.

Crews, David, and Michael C. Moore. "The Evolution of Brain Mechanisms Controlling Mating Behavior." *Science* 231 (1986): 121–25.

Crews, David, and Michael C. Moore. "Reproductive Psychobiology of Parthenogenetic Whiptail Lizards." In *Biology of Cnemidophorus Lizards*, ed. J. W. Wright. Seattle: University of Washington Press, in press.

Cuellar, Orlando. "Animal Parthenogenesis: A New Evolutionary-Ecological Model is Needed." *Science* 197 (1977): 837–43.

Cuellar, Orlando. "Further Aspects of Competition and Some Life History Traits of Coexisting Parthenogenetic and Bisexual Whiptail Lizards." In *Biology of Cnemidophorus Lizards*, ed. J. W. Wright. Seattle: University of Washington Press, in press.

Cuellar, Orlando. "Long-term Analysis of Reproductive Periodicity in the Lizard *Cnemidophorus uniparens*." *American Midland Naturalist* 105 (1981): 93–101.

Cuellar, Orlando. "Parthenogenetic Lizards [Response to Technical Comments]." *Science* 201 (1978): 1155.

Moore, Michael C., Joan M. Whittier, Allan J. Billy, and David Crews. "Male-like Behaviour in an All-female Lizard: Relationship to Ovarian Cycle." *Animal Behaviour* 33 (1985): 284–89.

Vanzolini, P. E. "Parthenogenetic Lizards [Technical Comment]." *Science* 201 (1978): 1152.

Wright, J. W. "Parthenogenetic Lizards [Technical Comment]." *Science* 201 (1978): 1152–54.

4. Texts Discussed: Chapter 5

Crews, David, and William R. Garstka. "The Ecological Physiology of a Garter Snake." *Scientific American* November 1982: 136–44.

Culliton, Barbara, revision of Jordan Gutterman et al. "Immunoglobulin on Tumor Cells and Tumor Induced Lymphocyte Blastogenesis in Human Acute Leukemia." *The New England Journal of Medicine* 288 (1973): 173–75.

Dunbar, Robin. "How to Listen to the Animals." *New Scientist* 13 June 1985.

Garstka, William R., and David Crews. "Female Sex Pheromone in the Skin and Circulation of a Garter Snake." *Science* 214 (1981): 681–83.

Gilbert, Lawrence. "The Coevolution of a Butterfly and a Vine." *Scientific American* August 1982: 110–21.

Ingelfinger, F. J. "Twin Bill on Tumor Immunity." *The New England Journal of Medicine* 289 (1973): 268–69.

"Leapin' Lizards: Lesbian Reptiles Act Like Males." *Time* 18 February 1981.

Massa, Don. "Working in Glass Houses." *Alcalde: UT Austin Alumni Magazine* January/February 1985: 6–8.

Mellanby, K. "Science Writing." *New Scientist* 15 March 1973: 625.

"Men and Women Under Forty Who Are Changing America : David Crews." *Esquire* December 1984: 36.

Papworth, M. "Oversimplified." *New Scientist* 12 April 1973: 123.

Parker, Geoffrey A. "The Reproductive Behaviour and the Nature of Sexual Selection in *Scatophaga stercoraria* L. (Diptera: Scatophagidae). IX. Spatial Distribution of Fertilization Rates and Evolution of Male Search Strategy Within the Reproductive Area." *Evolution* 28 (1974): 93–108.

Parker, Geoffrey. "Sex Around the Cow-pats." *New Scientist* 12 April 1979: 125–27.

Williams, Kathy S., and Lawrence E. Gilbert. "Insects as Selective Agents on Plant Vegetation Morphology: Egg Mimicry Reduces Egg Laying by Butterflies." *Science* 212 (1980): 467–69.

Wright, Pearce. "Deceiving Vine Keeps Butterflies at Bay." *The Times* (London) 20 August 1982.

5. Texts Discussed: Chapter 6

Allen, Elizabeth et al. "Against Sociobiology." *New York Review of Books* 13 (November 1976). Reprinted in *The Sociobiology Debate,* ed. Caplan. Pp. 259–64.

Alper, Joseph. "Ethical and Social Implications." In *Sociobiology and Human Nature,* ed. Gregory et al. Pp. 195–213.

Alper, Joseph, J. Beckwith, C. L. Chorover, et al. "The Implications of Sociobiology." In *The Sociobiology Debate,* ed. Caplan. Pp. 333–36.

Alper, Joseph, Jon Beckwith, and Lawrence G. Miller. "Sociobiology Is a Political Issue." In *The Sociobiology Debate,* ed. Caplan. Pp. 476–88.

Baerends, G. P. "Multiple Review of Wilson's *Sociobiology.*" *Animal Behaviour* 24 (1976): 700.

Barlow, George W. "Multiple Review of Wilson's *Sociobiology.*" *Animal Behaviour* 24 (1976): 700–701.

Bateson, Patrick. "Preface." In *Current Problems in Sociobiology,* ed. King's College Sociobiology Group. Cambridge: Cambridge University Press, 1982. Pp. ix–xi.

Beach, Frank. "Sociobiology and Interspecific Comparisons of Behavior." In *Sociobiology and Human Nature,* ed. Gregory et al. Pp. 116–35.

Burian, Richard M. "A Methodological Critique of Sociobiology." In *The Sociobiology Debate,* ed. Caplan. Pp. 376–95.

Campbell, Colin. "Anatomy of a Fierce Academic Feud." *New York Times* (November 9, 1986) section 12: 58–64.

Caplan, Arthur L. "Ethics, Evolution, and the Milk of Human Kindness." *Hastings Center Report,* April 1976. Reprinted in *The Sociobiology Debate,* ed. Caplan. Pp. 304–14.

Caplan, Arthur L., ed. *The Sociobiology Debate: Readings on Ethical and Scientific Issues.* New York: Harper and Row, 1978.

Crocker, Joe. "Sociobiology: The Capitalist Synthesis." *Radical Science Journal* 13 (1983): 55–72.

Dawkins, Richard. *The Selfish Gene.* Oxford: Oxford University Press, 1976.

Dunbar, R. I. M. "Adaptation, Fitness, and Evolutionary Tautology." *Current Problems in Sociobiology,* ed. King's College Sociobiology Group. Cambridge: Cambridge University Press, 1982. Pp. 9–28.

Dupree, A. Hunter. " 'Sociobiology' and the Natural Selection of Scientific Disciplines." *Minerva* (1977): 94–101.

Eckland, Bruce K. "Darwin Rides Again." In "Review Symposium." *American Journal of Sociology* 82 (1976): 692–697.

Gould, Stephen Jay. "Biological Potential vs. Biological Determinism." *Natural History Magazine* May 1976. Reprinted in *The Sociobiology Debate,* Caplan. Pp. 343–51.

Gregory, Michael S., Anita Silvers, and Diane Sutch, eds. *Sociobiology and Human Nature: An Interdisciplinary Critique and Defense.* San Francisco: Jossey-Bass, 1978.

Hammerton, M. "Rose's Faith." *New Scientist* 3 June 1976: 547.

Hinde, R. A. "Multiple Review of Wilson's *Sociobiology.*" *Animal Behaviour* 24 (1976): 706–7.

Hirsch, Jerry. "Multiple Review of Wilson's *Sociobiology.*" *Animal Behaviour* 24 (1976): 707–9.

Holton, Gerald. "The New Synthesis?" In *Sociology and Human Nature,* ed. Gregory et al. Pp. 75–97.

Hull, David L. "Scientific Bandwagon or Travelling Medicine Show?" In *Sociobiology and Human Nature,* ed. Gregory et al. Pp. 136–63.

King, James C. "The Genetics of Sociobiology." In *Sociobiology Examined,* ed. Montague. Pp. 82–107.

Krebs, John R. "Multiple Review of Wilson's *Sociobiology.*" *Animal Behaviour* 24 (1976) 709–10.

Krebs, John R. "Sociobiology Ten Years On." *New Scientist* 3 October 1985: 40–43.

Lewin, Roger. "The Course of a Controversy." *New Scientist* 13 May 1976: 344–45.

Midgely, Mary. "Rival Fatalisms: The Hollowness of the Sociobiology Debate." In *Sociobiology Examined,* ed. Montague. Pp. 15–38.

Miller, Lawrence. "Fated Genes." *Journal of the Behavioral Sciences* April 1976. In *The Sociobiology Debate,* ed. Caplan. Pp. 269–79.

Miller, Lawrence. "Philosophy, Dichotomies, and Sociobiology." *Hastings Center Report* 25 October 1976. Reprinted in *The Sociobiology Debate,* ed. Caplan. Pp. 319–24.

Montague, Ashley, ed. *Sociobiology Examined.* Oxford: Oxford University Press, 1980.

Maynard Smith, John. "The Birth of Sociobiology." *New Scientist* 26 September 1985: 48–50.

Parker, Geoffrey A. "Selfish Genes, Evolutionary Games, and the Adaptiveness of Behavior." *Nature* 274 (1978): 849–55.

Rose, Hilary, and Steven Rose. "Radical Science and Its Enemies." In *The Socialist Register 1979,* eds. R. Miliband and J. Saville. London: Merlin, 1979. Pp. 317–35.

Rose, Steven. "Sociobiology." *New Scientist* 20 May 1976: 433.

Rose, Steven. "Sociopolitics." *New Scientist* 17 June 1976: 663.

Rose, Steven. "It's Only Human Nature: The Sociobiologist's Fairyland." *Race and Class* 20 (1979). Reprinted in *Sociobiology Examined,* ed. Montague. Pp. 158–70.

Rose, Steven, R. C. Lewontin, and Leon J. Kamin. *Not In Our Genes: Biology, Ideology, and Human Nature.* Harmondsworth: Penguin, 1984.

Rosenblatt, J. S. "Multiple Review of Wilson's *Sociobiology.*" *Animal Behaviour* 24 (1976): 713–15.

Ruse, Michael. "Sociobiology: A Philosophical Analysis." In *The Sociobiology Debate,* ed. Caplan. Pp. 355–75.

Ruse, Michael. *Sociobiology: Sense or Nonsense?* Dordrecht: D. Reidel, 1979.

Sade, Donald Stone. "The Evolution of Sociality." In *The Sociobiology Debate,* ed. Caplan. Pp. 239–46.

Segerstrale, Ullica. "Colleagues in Conflict: An 'In Vivo' Analysis of the Sociobiology Controversy." *Biology and Philosophy* 1 (1986): 53–87.

Silcock, Brian. "How Genetic Is Human Behavior?" *Sunday Times* (London) 6 June 1976: 17.

Sociobiology Study Group. "Sociobiology: Another Biological Determinism." *Bioscience* 26 (1976). Reprinted in *The Sociobiology Debate,* ed. Caplan. Pp. 280–90.

Tinbergen, Niko. "On War and Peace in Animals and Man." (1968). Reprint in *The Sociobiology Debate,* ed. Caplan. Pp. 76–99.

Tobach, Ethel. "Multiple Review of Wilson's Sociobiology." *Animal Behaviour* 24 (1976): 712–13.
Tobach, Ethel. "The Methodology of Sociobiology from the Viewpoint of a Comparative Psychologist." In *The Sociobiology Debate*, ed. Caplan. Pp. 411–23.
Wade, Nicholas. "Sociobiology: Troubled Birth for a New Discipline." *Science* 191 (19 March 1976). Reprinted in *The Sociobiology Debate*, ed. Caplan. Pp. 325–32.
Waddington, C. H. "Mindless Societies." *New York Review of Books* 7 August 1975. Reprinted in *The Sociobiology Debate*, ed. Caplan. Pp. 252–58.
Washburn, S. L. "Animal Behavior and Social Anthropology." In *Sociobiology and Human Nature*, ed. Gregory et al. Pp. 53–74.
———. "Human Behavior and the Behavior of Other Animals." In *Sociology Examined*, ed. Montague. Pp. 254–82.
Wilson, E. O. "Academic Vigilantism and the Political Significance of Sociobiology." *Bioscience* 26 (1976). Reprinted in *The Sociobiology Debate*, ed. Caplan. Pp. 291–303.
Wilson, E. O. "Biology and the Social Sciences." *Daedalus* 106.4 (1977): 127–40.
Wilson, E. O. "For Sociobiology." *New York Review of Books* 11 December 1975. Reprinted in *The Sociobiology Debate*, ed. Caplan. Pp. 265–68.
Wilson, E. O. "Foreword." In *The Sociobiology Debate*, ed. Caplan. Pp. xi–xiv.
Wilson, E. O. "Human Decency Is Animal." *New York Times Magazine* 12 (October 1975): 39–50.
Wilson, E. O. "Introduction: What Is Sociobiology?" In *Sociobiology and Human Nature*, ed. Gregory et al. Pp. 1–12.
Wilson, E. O. "Multiple Review of Wilson's Sociobiology." *Animal Behaviour* 24 (1976): 698–99 and 716–18.
Wilson, E. O. *Sociobiology: The New Synthesis*. Cambridge: Harvard University Press, 1975.
Wynne-Edwards, V. C. "Bulls-eye of Sociality." *Nature* 259 (1975): 253–54.

6. Texts Discussed: Conclusion

Crook, J. H., and J. S. Gartlan. "Evolution of Primate Societies." *Nature* 210 (1966): 1200–1204. The table discussed is also given in E. O. Wilson, *Sociobiology*, p. 522.
"A Genetic Defense of the Free Market." *Business Week* 10 April 1978: 100.
Stein, Sara. *Girls and Boys: The Limits of Non-Sexist Childrearing*. London: Chatto and Windus, 1984.

Index